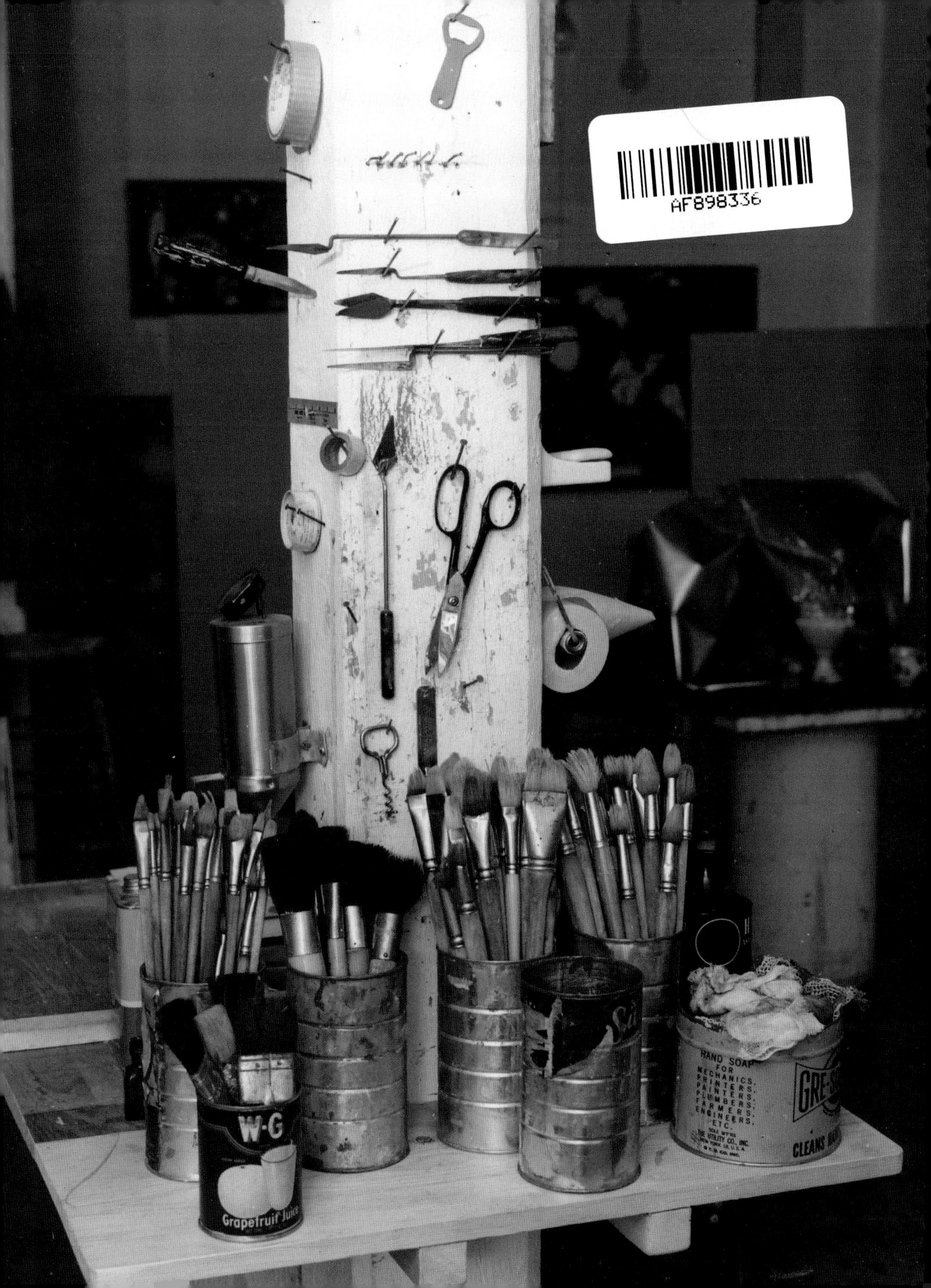
W·G
Grapefruit Juice
HAND SOAP
FOR
MECHANICS,
PRINTERS,
PAINTERS,
PLUMBERS,
FARMERS,
ENGINEERS,
ETC.

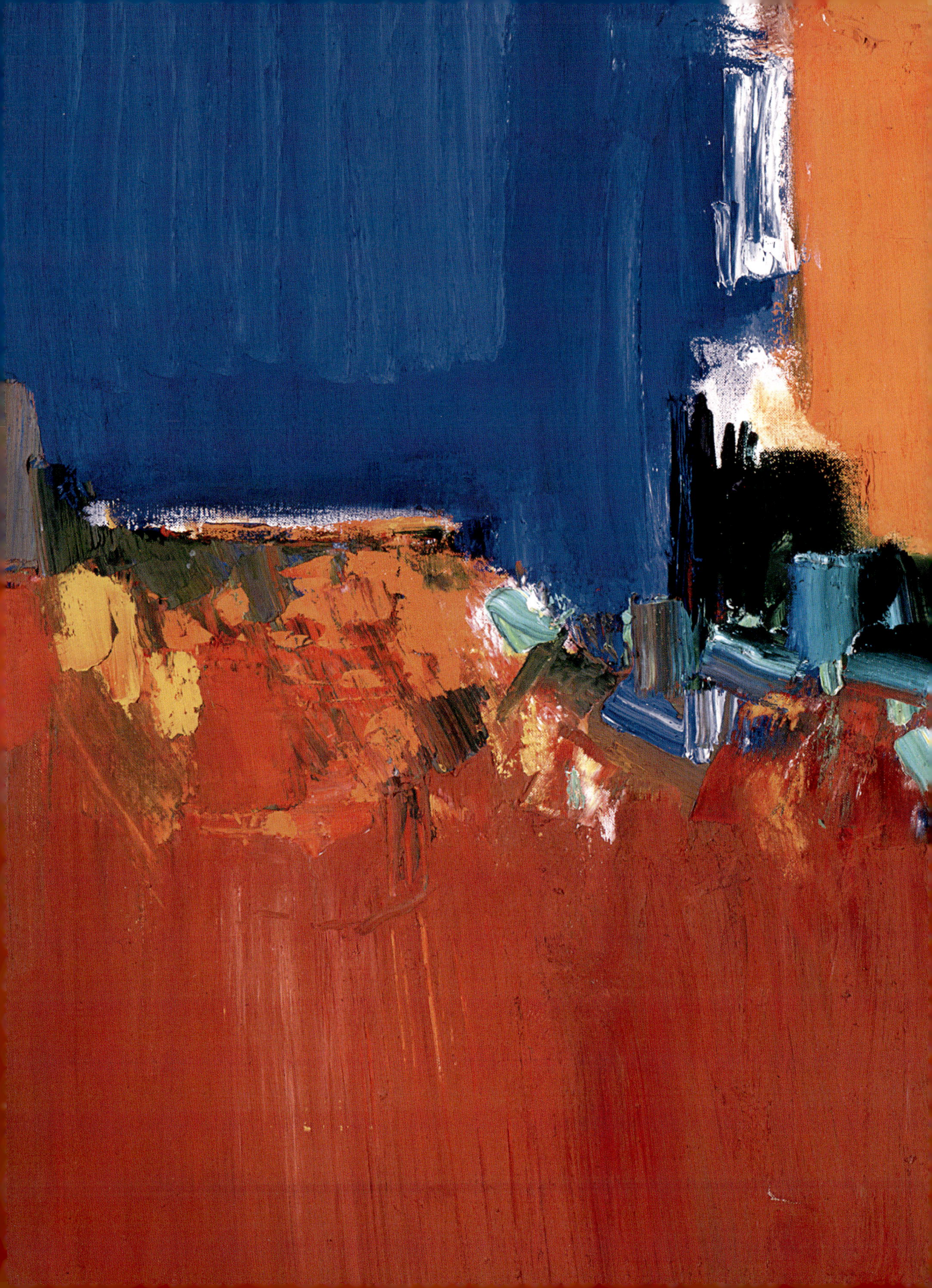

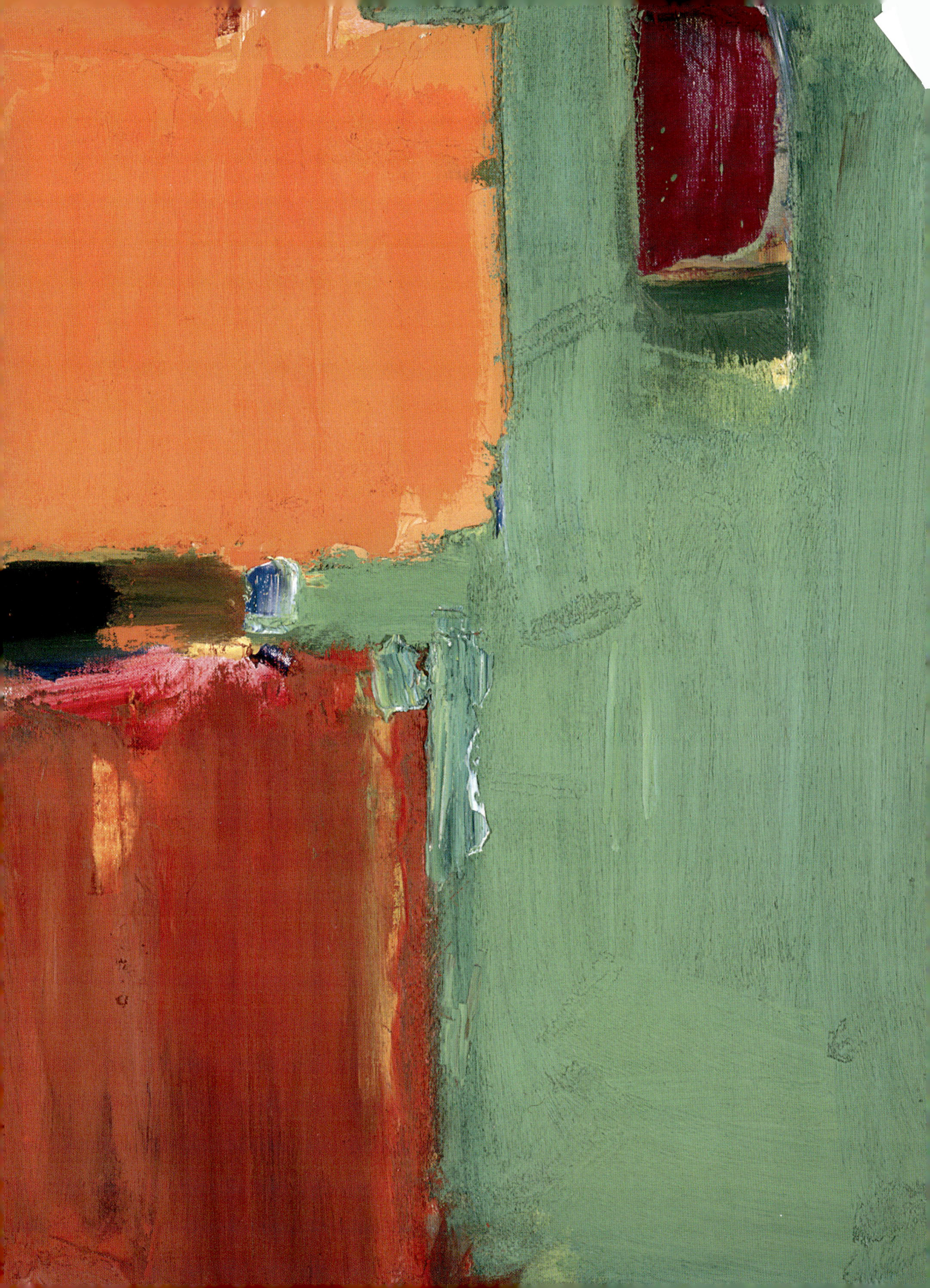

DAWN V. ROGALA

HANS HOFMANN
THE ARTIST'S MATERIALS

The Getty Conservation Institute
Los Angeles

Getty Conservation Institute

The Getty Conservation Institute works to advance conservation practice in the visual arts, broadly interpreted to include objects, collections, architecture, and sites. It serves the conservation community through scientific research, education and training, model field projects, and the broad dissemination of the results of both its own work and the work of others in the field. And in all its endeavors, it focuses on the creation and dissemination of knowledge that will benefit professionals and organizations responsible for the conservation of the world's cultural heritage.

Second printing

Published by the Getty Conservation Institute, Los Angeles
Getty Publications
1200 Getty Center Drive, Suite 500
Los Angeles, California 90049-1682
getty.edu/publications

Tevvy Ball, *Project Editor*
Jeffrey Cohen, *Designer*
Suzanne Watson, *Production*

Distributed in the United States and Canada by the University of Chicago Press
Distributed outside the United States and Canada by Yale University Press, London

FRONT COVER: Hans Hofmann, *Combinable Wall I and II*, 1961. Detail. See fig. 3.4.
FRONTISPIECE: Hofmann's New York studio, 1950. See fig. 3.16.
PAGES ii AND iii: Hans Hofmann, *Indian Summer*, 1959. Detail. See fig. 3.7.
BACK COVER: Hofmann at work, 1950. Detail. See fig. 3.17.

Names: Rogala, Dawn V., author. | Hofmann, Hans, 1880–1966, artist. | Getty Conservation Institute, issuing body.
Title: Hans Hofmann : the artist's materials / Dawn V. Rogala.
Description: Los Angeles : Getty Conservation Institute, [2016] | Includes bibliographical references and index.
Identifiers: LCCN 2016009233 | ISBN 9781606064870 (pbk.)
Subjects: LCSH: Hofmann, Hans, 1880–1966. | Hofmann, Hans, 1880–1966—Criticism and interpretation. | Painting, Modern—20th century—Technique. | Artists' materials—Analysis. | Paint materials—Analysis.
Classification: LCC ND237.H667 R64 2016 | DDC 759.13—dc23
LC record available at https://lccn.loc.gov/2016009233

Contents

Foreword

Hans Hofmann, one of the key figures in postwar American art, is justly known for playing a pivotal role in the development of the Abstract Expressionist movement. Throughout his career, from early twentieth-century Munich and Paris to the seminal epoch of American modernism in midcentury New York, he was widely respected as a teacher as well as a painter.

Hans Hofmann: The Artist's Materials is the fourth book in the Getty Conservation Institute's Artist's Materials series, which publishes in-depth research into the materials and techniques of significant twentieth- and twenty-first-century artists. The series grew out of the GCI's Modern and Contemporary Art Research initiative, conceived to explore more fully the questions and challenges of conserving modern and contemporary art. This period has seen a staggering expansion in the types of materials used by artists, and knowledge of these materials and the techniques used in working with them is sorely needed by conservation professionals. Many conservators and researchers are currently carrying out studies on these materials—often in collaboration with the GCI—and the Artist's Materials books are an effective way to disseminate the results of their research. Previous books in the series discuss Willem de Kooning (2010, second printing 2011), Jean Paul Riopelle (2011), and Lucio Fontana (2012).

The study of Hofmann's work illuminates a range of both stylistic and material innovations related to this seminal period of modern art. The author of this book, Dawn Rogala, is a paintings conservator at the Museum Conservation Institute, Smithsonian Institution. She received her PhD in preservation studies from the University of Delaware, where her research involved examining Hofmann's works and identifying his late-career materials, using paintings primarily from the University of California, Berkeley Art Museum and Pacific Film Archive, which holds the largest public collection of his work. We thank Dawn for her work on this informative and handsome volume.

We trust that this book will engage a wide audience, including conservators, conservation scientists, students, art historians, curators, and others who enjoy and care for the work of Hans Hofmann and his contemporaries.

Timothy P. Whalen
Director
The Getty Conservation Institute

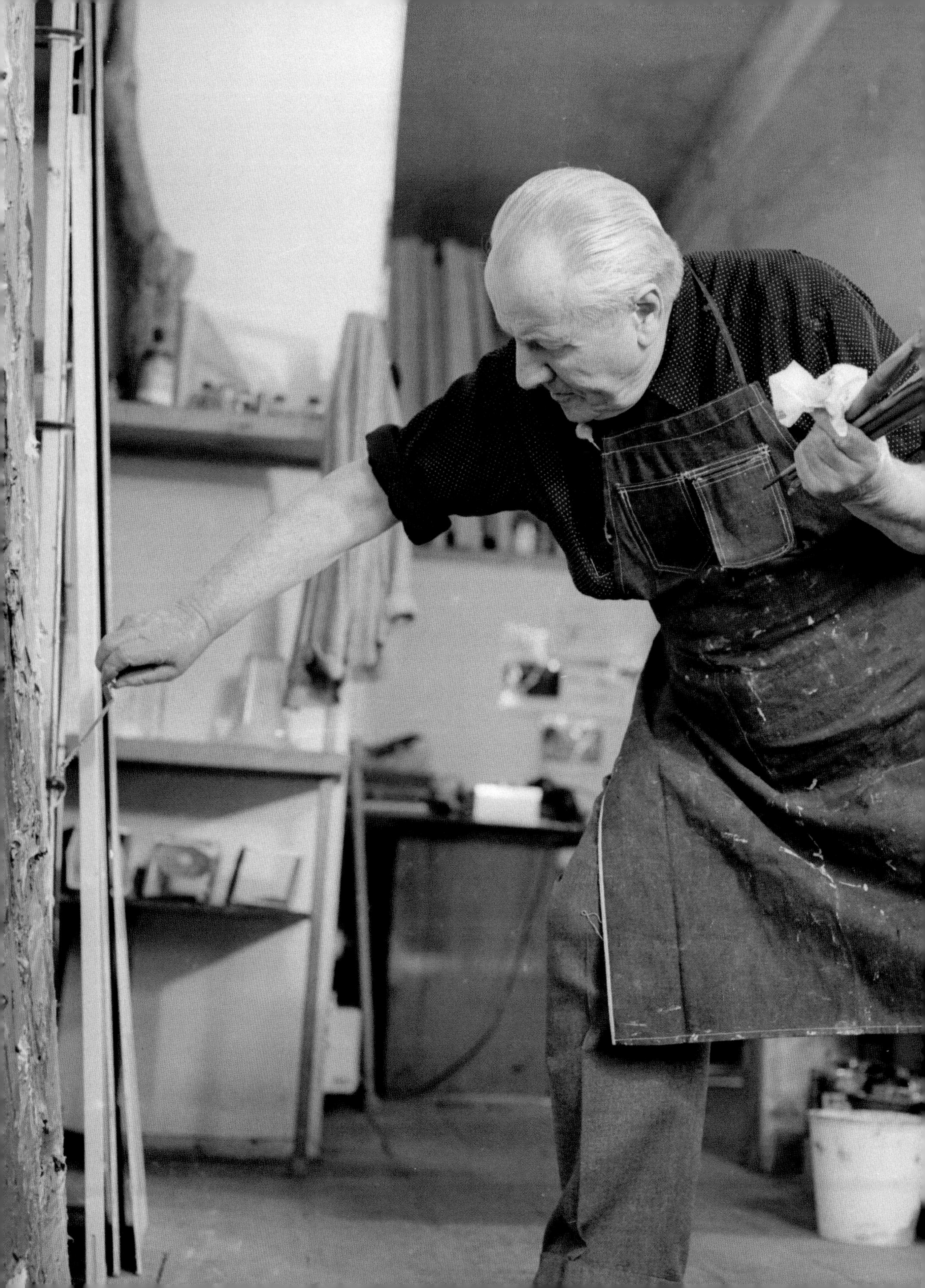

Introduction

> What you're going to hear tonight is a defense of Abstract Expressionism, [and] if there is to be a defense of Abstract Expressionism at all... it will have to be cast as a defense of Hofmann in particular.[1]
>
> —T. J. CLARK, 1992 lecture at the University of California, Berkeley

The career of the influential painter and educator Hans Hofmann (1880–1966) describes the arc of artistic modernism from its European cradle in pre–World War I Munich and Paris to the lofts and communal art centers of mid-twentieth-century Greenwich Village. His career traces, too, the transatlantic engagement of modern painting with the materials of its own making. In these interrelated narratives, Hofmann was a central protagonist. Hofmann provided a vital link in the continuum between nineteenth- and twentieth-century arts practice and between European and American modernism, and his stature merits comparison with that of more widely known figures, such as Pablo Picasso, Willem de Kooning, and Jackson Pollock. Hofmann's authority with practitioners and critics alike positioned his schools (run uninterrupted from 1915 to 1958) as a primary destination for advanced arts practice and theory and made Hofmann simultaneously the influencer and the benefactor of the leading edge of modern art practice. The remarkable vitality of the artist's late-career work, produced in an atmosphere of unparalleled access to experimental painting theory and technique, affords insight not only into the style but also the literal substance of this formative period of artistic and material innovation. Indeed, Hofmann's signature late-career paintings tie together the efforts of his progenitors, colleagues, and descendants—"which is to say," in the words of the artist Frank Stella, "all of the twentieth century" (Stella 1999, 16).

Hofmann's late works also show us why and how we need to change our thinking about the conservation of modern art. Hofmann's materials and technique present an example of period artistic method that is underrepresented in conservation research but has direct implications for the ways in which conservators should treat Abstract Expressionist and later works. The embrace of new materials and processes is one of the great legacies of Abstract Expressionism, but the idea that these twentieth-century modernist innovators wholly abandoned traditional painting practice is a myth that works against the preservation of that era's artistic legacy. A mix of old and new is more representative of period artistic method than is suggested by the nascent library of technical studies focused on new materials or procedures. In fact, many of the condition problems art conservators face in the treatment of modern paintings are directly related to a mixing of familiar techniques and modern materials, an approach apparent in Hofmann's work and

Hans Hofmann, 1957.
Photo: Andreas Feininger

that of his contemporaries. A thorough examination of Hofmann's late-career materials and techniques allows us to reassess the critical view of Abstract Expressionist painting as dependent on materials; understanding the transitional nature of art practice during this period informs the future conservation treatment of these and later works that incorporate both traditional and modern paint media.

Technical scholarship and aesthetic inquiry are not mutually exclusive. As modern materials research has advanced, so too has the call for an inclusive model of scholarship that acknowledges the advantages of interrelated conservation and humanities studies. "The two disciplines are complementary," noted researchers at the Netherlands Cultural Heritage Agency. "The material-technical expertise of the conservator/restorer and the curator's art-historical and theoretical knowledge are the foundation upon which we base an understanding of the art work" (Hummelen et al. 1999, 315). Technical studies do not ignore the creative act; they are vital to its full interpretation. Providing the data is one part of the process, according to Carol Mancusi-Ungaro, associate director for conservation and research at the Whitney Museum of American Art and founding director of the Center for the Technical Study of Modern Art, "but explaining it in context is quite another.... It is only through intense collaboration among the distinct but related disciplines that...we begin to offer the indeterminate work of art the rigorous yet insightful review it deserves" (2005, 161). Hofmann's unprecedented access to the leading edge of international arts practice, along with his influential role within the American modernist community, make an interdisciplinary study of his paintings a particularly appropriate and informative addition to Abstract Expressionist scholarship.

This book focuses primarily on the last decade of the artist's production (1955 through 1965), a period that includes the years just prior to and after the 1958 closing of Hofmann's art schools in New York City and Provincetown, Massachusetts, and coincides with the emergence of the artist's signature style. A representative catalogue of Hofmann's late-career materials has been built from my analysis of over five hundred paint and fiber samples from palettes and works on canvas produced from the late 1940s until the artist's death in 1966, and the materials identified through this analysis are compared with existing conservation survey and treatment documentation of Hofmann's work and with both published and anecdotal assessments of his rapid stylistic changes during this period. This research defines the limited but important role played by new paint media in Hofmann's late-career style and clarifies the impact of mixed traditional and modern painting materials in the study and preservation of Abstract Expressionist paintings. The information provided in this book expands upon my previous research into the relationship between Abstract Expressionist materials and the physical and aging characteristics of Abstract Expressionist paintings, revealing similarities in materials used by Hofmann and his New York School colleagues and allowing us to reevaluate methodologies for the conservation of Abstract Expressionist and later works that incorporate contemporary industrial binders and synthetic pigment formulations into otherwise traditional artists' palettes.

In this book I analyze and identify the materials used by Hofmann in his late-career paintings, assess notable relationships between his materials and technique, and examine the impact of his choices on the long-term stability of his work. Twenty-six paintings and eight palettes were selected for analysis as exemplars, including twenty-three paintings on canvas from the years 1953 through 1965. Most of these paintings are from the Hofmann Collection at the University of California, Berkeley Art Museum and Pacific Film Archive—the world's largest and most comprehensive public collection of the artist's work. These are supplemented by works from the Albright-Knox Art Gallery, Buffalo, New York; the Memorial Art Gallery, Rochester, New York; and New York's

Museum of Modern Art. Palettes were selected from the holdings of the Renate, Hans and Maria Hofmann Trust, New York. Analysis, which aimed at the broad characterization of materials, had two primary goals: first, to confirm and track Hofmann's use of industrial materials in his ground layers, and second, to identify and track his use of any newly developed pigments or paint media in his compositional layers.

Information regarding ground layer stratigraphy and identification of inorganic pigments was obtained with optical microscopy and scanning electron microscopy–energy dispersive spectroscopy (SEM-EDS). Fourier transform infrared spectroscopy (FTIR) and gas chromatography–mass spectrometry (GC-MS) were used to identify binders and organic pigments, and X-ray diffraction (XRD) was employed in a limited capacity to assess similar pigment blends. Materials analysis was performed at the scientific laboratories of the National Gallery of Art, Washington, DC, and the Museum Conservation Institute of the Smithsonian Institution.[2] Primary sources of information related to the study group paintings include the conservation records of the Hofmann Collection held at the Berkeley Art Museum and Pacific Film Archive and the San Francisco Museum of Modern Art, and two unpublished surveys of the collection's condition and treatment history compiled by conservators at the San Francisco Museum of Modern Art and the Intermuseum Conservation Association in Oberlin, Ohio (now located in Cleveland).

Archival resources utilized for first-person accounts of Hofmann's materials and methods include the oral history records of the Smithsonian Institution's Archives of American Art and the University of California Regional Oral History Office, as well as the Hofmann-related collections housed at the Archives of American Art and the Bancroft Library of the University of California, Berkeley.

This book integrates recent scholarship in art history, conservation, and scientific analysis while defining the relationship among Hofmann's materials, his work, and the preservation of his artistic legacy. It is organized to provide context and clarity for readers from related disciplines as well as for general readers with an interest in the various fields of art and art historical study. Chapter 1 traces Hofmann's exposure to the avant-garde arts communities of late nineteenth and early twentieth-century Europe and follows him to the United States, while Chapter 2 situates the artist and his seminal Hans Hofmann School of Fine Arts within the Abstract Expressionist community in America. These early chapters also illuminate the role of materials in Abstract Expressionist art, emphasizing the parallel innovations in modern painting and modern paint technology, revealing the impact of early twentieth-century developments in paint materials on modern art movements, and elucidating the continuing modernist embrace of these new materials. The book's third chapter presents Hofmann's paintings, examines their material nature, and evaluates the trajectory of his late-career style while also considering the role of materials in the artist's aesthetic philosophy. Chapter 4 covers the selection of artworks for the technical study, discusses behaviors unique to Hofmann's work and related behaviors common to modern paintings, and reviews past and current conservation methodology in the paintings' treatment. The final two chapters present the results of the materials analysis and discuss Hofmann's late-career materials in relation to the modernist communities within which Hofmann lived in Europe and the United States, thereby positing a new understanding of Abstract Expressionist materials in general. We conclude with a call for an inclusive technical scholarship on Abstract Expressionism that acknowledges the ramifications of transitional arts practice and the role of physical evidence in art historical study and emphasizes their implications for conservation methodology.

In tracing Hofmann's relationship to his materials, this book also seeks to offer a new perspective on the role of Abstract Expressionism within the larger modernist continuum. The works of Hofmann and his contemporaries distilled the advances of preceding generations and directly shaped the work of their modernist descendants. Through Abstract Expressionism, direct engagement with the materials of painting emerged as a singular influence on artists, critics, and audiences, and the full impact of this determinative movement is best understood by looking at the period through an informed eye positioned at the nexus of activity. Hofmann is that informed eye, placed at the pivotal moment, and the study of his work enhances our understanding of modern art history. "Hofmann's art," noted *New York Times* critic and former Museum of Modern Art curator Sam Hunter, "is surely one of the remarkable examples in the century of modern artistic style of consciousness reporting objectively on itself" (Hunter 2002, 31).

1 A Life in Art

A painting…is an act inseparable from the biography of the artist. The painting is a "moment" in the adulterated mixture of his life.
—HAROLD ROSENBERG, "The American Action Painters," 1952

You cannot help it. You belong to a certain time. You are yourself the result of this time. You are also the creator of this time.
— HANS HOFMANN, 1966

Johann (Hans) Georg Albert Hofmann was born on March 21, 1880, the second of five children of Franziska Manger and Theodor Hofmann, who lived in the Bavarian town of Weissenburg, where Hofmann's father worked as a city clerk. The family moved at least once before settling in Munich in 1886; there Hofmann's father took a position as a district clerk with the Bavarian Ministry of the Interior (Dickey 2011, 33–34).[3] There is little information about Hofmann's artistic beginnings, a matter complicated by anecdotal recollection and Hofmann's flair for self-promotion. A biographical account included in the catalogue for Hofmann's 1948 exhibition at the Addison Gallery of American Art, in Andover, Massachusetts, for example, was fashioned by gallery director Bartlett H. Hayes Jr. as a parable focused on atmosphere "rather than true chronological circumstances" (Hofmann 1948, xiv). According to Hayes, Hofmann liked the story and insisted that it be part of the catalogue. The young Hofmann is said to have excelled in music and science, and Hofmann told historian and painter Frederick S. Wight that he had applied for a patent—through his mother, because he was underage—for an "electromagnetic comptometer" (Wight 1957, 26), although no records of the patent have been found. One oft-repeated story reports that at the age of sixteen Hofmann received a thousand-mark gift from his father to pursue his engineering interests but instead used the money to enroll in art classes in Munich's Schwabing arts district (Dickey 2011, 35), the center of the German avant-garde.

The Young Artist in Germany

Hofmann was positioned at the right historical place and time to study modern painting. Many formative influences on the aesthetic and compositional philosophies of Hofmann and his Abstract Expressionist colleagues can be traced to turn-of-the-century Munich, which, along with Paris, was one of the most important modernist centers before World War I. Despite growing pressure from conservatives to suppress nontraditional art they considered decadent (Makela 1996), the Munich of Hofmann's youth was at the forefront of modern aesthetics—it was a city with "an artistic life of extraordinary activity and variety," according to art historian Peter Selz (1997, 173). In the final years of the nineteenth century, University of Munich professor Theodor Lipps lectured on the relationship between artist and subject, while Heinrich Wölfflin completed his dissertation on the psychological response to architectural forms. Hofmann's early years as an artist also coincided with Munich sculptor Adolf Hildebrand's influential 1893 treatise *Das Problem der Form in der Bildenden Kunst* (*The Problem of Form in Painting and Sculpture*), in which he laid out the concepts of planar depth and motion that influenced artists and critics alike. "Hildebrand's [text] was epoch-making in its effect on art criticism, which entered a new era with its publication," noted Selz (1997, 6). "Art education, art history and the artist [had] been merely ignorant about space as the predominant problem in the visual arts," Hofmann would later write to one of Meyer Schapiro's graduate students, "up to the appearance of Adolf Hildebrand's book" (Dickey 1997, 83).[4]

Students with an interest in modern art had ample options for study, as a rising number of art schools appeared in Munich in response to German population growth (McClelland 1996). The new schools transformed the Schwabing neighborhood surrounding the Akademie der Bildenden Künste München—the Munich Academy of Fine Arts—and the artists who took advantage of Schwabing's concentration of studios and who studied with multiple teachers were exposed to a range of modernist styles and influences. During his early years studying in Schwabing, Hofmann built lasting friendships with such fellow artists as painter and classmate Jules Pascin and poet Rainer Maria Rilke, and there he met Maria ("Miz") Wolfegg, his future wife. A portrait of Miz, among the few paintings remaining from Hofmann's early studies, shows the prominent influence of such modernists as Pierre-Auguste Renoir and the "dark impressionist manner" noted by Hunter (1964, 11) as popular among Munich artists (fig. 1.1).

Hofmann studied with several teachers around Munich. "The names of his teachers are dim now," recounted Wight. "Michailow, Aspe [likely Anton Ažbe], the Hungarian painter Ferenczi, and Grimwal. . . .Willie Schwarz, . . . had been in Paris and . . . came back full of a new discovery—impressionism" (1957, 28).[5] Impressionist and Post-Impressionist theory were also on view in Munich in such venues as the Secession Gallery, which exhibited the work of local avant-garde artists alongside the progressive aesthetics of their French counterparts.

While studying modern painting technique under Schwarz at Moritz Heymann's studio, Hofmann learned the theories of contrasting and adjacent color disseminated by such scientists as Michel-Eugène Chevreul and Ogden Nicholas Rood and explored in the paintings and writings of French artists including Georges Seurat and Paul Signac—whose influences can be seen in an extant self-portrait of Hofmann from 1902 (fig. 1.2). When studying with former Academy of Fine Arts professor Anton Ažbe, Hofmann was introduced to the optical science of Munich physicist Hermann Ludwig Ferdinand von Helmholtz. "Helmholtz's studies on the physiology and psychology of color vision," according to Selz (1997, 3), "acted upon much of German artistic production" of the period. Ažbe's school attracted students interested in learning the Slovenian painter's own technique for manipulating adjacent color effects—a practice Ažbe called color crystallization, or *Farbkristallisation* (Dickey 2011, 38). Ažbe's popular technique drew a large contingent of modernist students to the teacher's 16 Georgenstrasse studio, including nonobjective painting pioneer Wassily Kandinsky, whose classmates included fellow Russian modernist Alexei Jawlensky (Selz 1997, 176), with whom Kandinsky later founded the first modernist secession group in Munich—Neue Künstlervereinigung

Modern Paint Technology in Germany

Hofmann began his modern painting studies in a period —and region—of advancing paint technology. In the late nineteenth century, Germany was a leader in the development of coatings. Following the 1856 discovery of aniline purple dyes by German chemist August Wilhelm Hofmann,[a] the founding of chemical manufacturers Badische Anilin- & Soda-Fabrik (BASF) and Friedrich Bayer (Friedr. Bayer & Co., later Bayer AG) made Germany a leading center for color research and development. While Hofmann was studying modern color theory in Munich, German manufacturers added a range of new organic and inorganic color options to artists' palettes.[b] Contemporary painters had access to bright zinc whites and bold cadmium yellows and oranges; synthetic organic colors built from red and purple alizarin and rhodamine dyes; and the range of azo, diazo, and fluorescein yellows and oranges (Boxall 1978a; Lomax and Learner 2006). German-manufactured blues, greens, and cadmium yellows were embraced by progressive painters across Europe, including Cézanne and Monet.

As a member of the Munich arts community, Hofmann had direct exposure to dramatic advances in the development of paint technology and the related growth of the artists' materials industry (B. Miller 1998). According to sociologist Janet Marontate, "There was a strong tradition of interest in the study of art materials in Germany and there were important precedents for the collaboration of artists, paintmakers and other interested parties" (1996, 267). In 1881 chemists Adolf Wilhelm Keim and Ernst Berger installed their Versuchsanstalt für Maltechnik (Research Institute for Painting Techniques) at the Munich Academy as a "laboratory and information office for painting techniques"—where, three years later, Keim produced a workable version of his potassium silicate paint. Keim's lectures on "mineral painting" were thereafter added to the Academy curriculum (Kinseher 2006, 42). Along with painter Franz von Lenbach and chemist Max Joseph von Pettenkofer, Keim was instrumental in the 1886 formation of the Deutsche Gesellschaft zur Beförderung rationeller Malverfahren (German Society for the Promotion of Rational Painting Techniques). Established to include members from artistic, industrial, and scientific circles in Munich, the society was created, according to conservator Kathrin Kinseher (2006, 41), to "counteract the lack of objective technical knowledge regarding the production of durable paints" and to create a network of professors, painters, scientists, and manufacturers in "an alliance of interests." Because the society's founders felt that an institutional infrastructure for research into painting materials and techniques and the durability of artworks was lacking, in 1892—fifty years before similar efforts in the United States—members of the society petitioned the German government to establish national standards for the testing and labeling of artists' paints and began efforts to educate the community on art-making materials.

"German artists have for many years taken an active interest in artists' materials," society member and Munich Academy professor Max Doerner (or Dörner) noted in a 1921 publication of his compiled lectures, "and their activities have resulted in advantages to all artists" (Doerner 1984, 94–95). In 1893 the Munich Glaspalast was chosen as host to the society-sponsored Exhibition of Painting Techniques, an international exposition of the latest advances in paint techniques and technology. The journal *Technische Mitteilungen für Malerei* (*Technical Instructions for Painting*) called upon "artists, decorators, manufacturers, inventors, and material, art and teaching-aid suppliers" to contribute to the sampling of traditional and experimental paint media and tools and the presentation of experimental paint techniques. Along with representatives from chemical manufacturers and technical journals, over forty German textile mills, paint manufacturers, and brush suppliers participated in the exhibition (Kinseher 2006, 41–44). The event underscored Munich's position as the center of advanced painting and paint making.

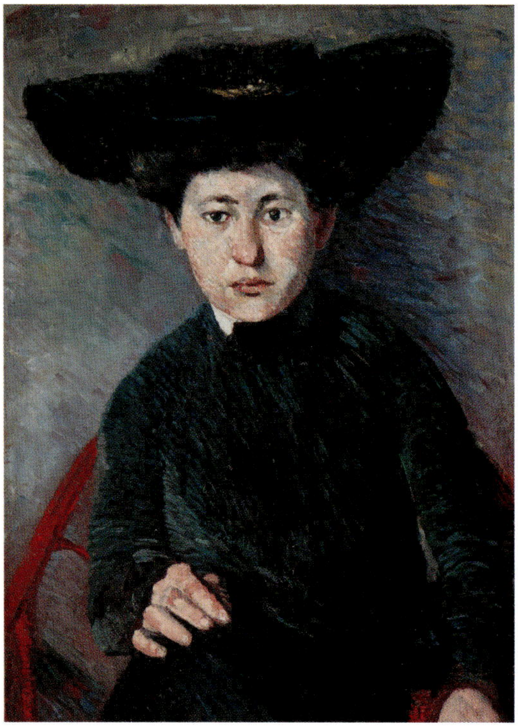

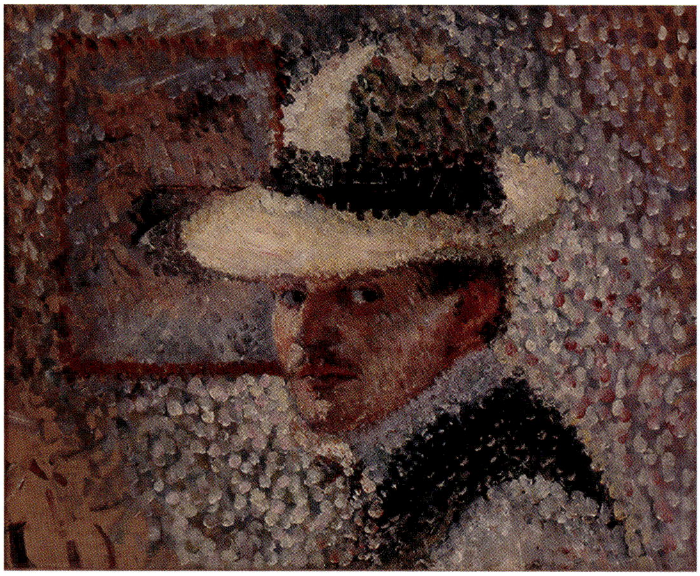

FIGURE 1.1
Hans Hofmann, *Maria Wolfegg, Portrait*, [1901]. Oil on board; 69.9 × 49.8 cm (27½ × 19⅝ in.). Munich, Städtische Galerie im Lenbachhaus, G18054. Gift of the Renate, Hans and Maria Hofmann Trust.

FIGURE 1.2
Hans Hofmann, *Self-Portrait*, 1902. Oil on board, 42.2 × 51.4 cm (16⅝ × 20¼ in.); painting on verso. New York, private collection.

München (also known as NKVM, or the New Munich Artist's Association)—whose annual exhibitions at the Moderne Galerie in Munich showed work by French and Russian group members, including avant-garde artists Pablo Picasso and Georges Braque and members of the NKVM splinter group Der Blaue Reiter (The Blue Rider).[6] "Abstract Expressionism was and is just as European," claimed writer Michel Seuphor (the pseudonym of Belgian painter Fernand Berckelaers), "since it was born in Munich" (Harrison 2007, 13).

The Paris Years

In 1904, Berlin art dealer Paul Cassirer—who had seen Hofmann's work at the request of Schwarz—secured for Hofmann the patronage of German department store magnate Philipp Freudenberg.[7] Freudenberg's financial support facilitated Hofmann's move to Paris. Hofmann's arrival in the French capital coincided with an influx of modern painters around the time of the Salon d'Automne, which that year featured an entire room dedicated to the work of Paul Cézanne and the following year hosted the inaugural exhibition of bold works by artists soon after referred to as the Fauves.[8] The 1904 salon also featured the work of the influential young colorist Robert Delaunay, with whom Hofmann was to form a strong friendship. A few months later Hofmann joined Wolfegg in Belgium (Wight 1957, 28), and they journeyed together to Paris, where they stayed for the next ten years, first at artists' housing built on the rue Campagne Première from materials salvaged from the 1889 Exposition Universelle (Dickey 2011, 40) and later at lodgings on the rue de Sèvres.

Paris was the center of the modern art world at the outset of the twentieth century. Students flocked to Parisian ateliers in search of training, and as in Munich, Paris responded to the increase in students with an increase in private art schools, including Henri Matisse's short-lived but well-known school at the Couvent des Oiseaux, near Mont-

Modern Paint Technology in France

Paint technology advanced as dynamically as aesthetics during Hofmann's years in Paris, and progressive French painters already familiar with the color theories of Michel-Eugène Chevreul and Ogden Nicholas Rood had likewise embraced innovative paint formulations. Synthetic pigments were commonplace on the Post-Impressionist and Pointillist palettes of such artists as Vincent van Gogh and Georges Seurat (Kirby et al. 2003; Van Bommel, Geldof, and Hendricks 2005); according to science writer Philip Ball, of the twenty principal pigments identified by analysis in Impressionist paintings, twelve were recent synthetic formulations—primarily deep shades of blue and green (Ball 2001a, 181; 2001b). According to research scientist Matthijs de Keijzer (2002, 42), "[today's] pigment lists of the well-known paint manufacturers such as Royal Talens [Netherlands], Winsor & Newton [England], Daler-Rowney [England], Schmincke [Germany], Lukas [Germany] and Société Lefranc & Bourgeois [France] show that...half of the quality A artists' paints contain one or more organic and inorganic compounds discovered in the 20th century."[c]

The continuous innovation in paint technology that took place during Hofmann's years in Paris made possible the striking shift in palette seen in the avant-garde paintings of the Fauves and members of Der Blaue Reiter and in the new, modern color that served a primary creative function in Orphism and Synchromism. Such artists as Matisse, Delaunay, and André Derain embraced the expanding range of bold Hansa (arylide) yellows[d] and the cadmium colors shunned by more established colleagues, including Impressionist Pierre-Auguste Renoir, who, according to art historian John Gage, "refused to try even the sample of cadmium Matisse gave him, saying that he did not want to change his ways" (Ball 2001a, 303).[e]

Arylide enamel colors found in the paintings of Picasso and Vaclav Vytlacil made use of both the new colors and the oil-modified alkyd polymer resins that began to appear in paint formulations in 1911 (Boxall 1978a, 1978b; Lomax and Learner 2006, 109). These innovative materials were also featured three years later at an exhibition of commercial paints and dye-based pigments sponsored by the Munich-based industrial advocate group Deutsche Werkbund and organized by chemist Wilhelm Ostwald, a consultant to the German paint industry and author of the 1904 book *Malerbriefe: Beiträge zur Theorie und Praxis der Malerei* (*Letters to a Painter: Contributions to the Theory and Practice of Painting*).

The activities of the Munich-based Deutsche Gesellschaft zur Beförderung rationeller Malverfahren during this period included a 1905 conference on paint adulteration organized in part by Max Doerner, who served as the organization's president from 1910 through 1913. Doerner included the organization's list of recommended artists' colors (Normalfarben) in the compiled lecture notes on artists' health and safety practices that would become his influential 1921 treatise *The Materials of the Artist and Their Use in Painting* (Doerner 1984).

While Doerner suggested that early commercial aniline pigments were not stable and deemed their commercial release "precipitous and ill-advised," he considered other modern formulations to be improvements on traditional materials and encouraged all artists to maintain an open yet discerning attitude toward new paint media. "The laws which govern the materials of the artist," Doerner stated, "are the same for all artists, to whatever schools they may belong" (1984, 90, v).

parnasse, and the Académie Julian, within the Passage des Panorama in Montmartre.[9] In his early days in Paris, Hofmann trained at ateliers along rue de la Grande Chaumière and participated in classes at both the Académie de la Grand Chaumière and the neighboring Académie Colarossi,[10] whose evening class attracted notable students, including Matisse and the American modernists Lyonel Feininger and Max Weber. "France has fertilized the ideas of the whole world," Hofmann later declared, most notably in a pamphlet distributed at a 1949 panel discussion entitled "French Art vs. U.S. Art Today" (fig. 1.3) and in a subsequent interview for *Arts and Architecture* magazine:

> The greatness of Paris is to offer an invitation of participation to the spirit.... Paris' humanism has given the opportunity for free development of such artists as the Spaniards, Picasso, [Juan] Gris, [Joan] Miró; the Swiss, [Paul] Klee, [Jean] Arp, [Le] Corbusier, [Alberto] Giacometti; the Russians, Kandinsky, [Marc] Chagall; the Rumanian, [Constantin] Brancusi; the Irishman, [James] Joyce; the Americans, [Gertrude] Stein, Man Ray, [Arthur B.] Carles; the Pole, [Jacques] Lipschitz [Lipchitz]; and many others. Still today many of them are living masters of Paris. (Ellsworth 1949, 45)

There are several accounts of Hofmann's close connections to the art elite during his years in Paris (Goodman 1986, 18; Dickey 2011, 44), including a chance conversation between Leo Stein and painter Glenn Wessels as Wessels accompanied Hofmann during his first trip to the United States. Stein told Wessels that "Hofmann had been one of... the same group that was made up of pretty near all the early Cubists, Picasso, Delaunay, [Juan] Gris, [Francis] Picabia, Matisse, all these men, and had known them, rubbed shoulders with them, and had argued with them" (Wessels 1967, 117) (fig. 1.4). Hofmann often claimed to have painted on the same hotel balcony where Matisse painted his well-known views of the Seine (Wight 1957, 29), and in fact, it was Matisse's enthusiastic reaction to an exhibition of Hofmann's work at Cassirer's gallery that secured the continued patronage of Freudenberg during Hofmann's decade in France (Wight 1957, 30–31; Friedel 1997, 8).[11]

Hofmann's years in Paris also brought him into the social circles of other influential art dealers, including painter and dealer Richard Goetz (Goodman 1986, 18) and art dealer Wilhelm Uhde. Hofmann may have met Uhde through Jules Pascin—Hofmann's friend from Moritz Heymann's school in Munich—and during the Hofmanns' years in Paris, Wolfegg was friends with Pascin's wife, Hermionette David, and with Uhde's wife, the painter and designer Sonia Terk (Goodman 1982, 21; Dickey 2011, 44), who later married Delaunay. As with Hofmann's early years in Munich, the artist's time in the French capital is not clearly accounted for. According to critic and curator Sam Hunter, once Hofmann arrived in Paris, he had no contact with German avant-garde artists until the 1920s (1964, 11),[12] but this is unlikely, considering Hofmann's association with German dealers Goetz and Uhde and his close ties to Terk (who was German) and Delaunay. Delaunay and Terk were members of Der Blaue Reiter—whose founding members included Kandinsky, Franz Marc, and August Macke—and visits to the Delaunays' studio on the rue des Grands Augustins would have kept Hofmann in contact with members of the German avant-garde and possibly introduced him to such like-minded American modernists as Morgan Russell and Stanton Macdonald-Wright, whose abstract "synchromies" were on view in both Munich and Paris.[13] Hofmann even exhibited with the Berlin Secession group in 1908 and 1909 and shared a joint exhibition at Cassirer's gallery with Austrian painter Oskar Kokoschka.

FIGURE 1.3
Illustration from the *Forum '49* pamphlet, reproduced in the November 1949 issue of *Arts and Architecture*. Hans Hofmann, *Untitled*, 1949. Ink and gouache on paper, 30.5 × 48.3 cm (12 × 19 in.). Gift of the Renate, Hans and Maria Hofmann Trust, M-0631/04.

Photo: Thomas Quigley.

FIGURE 1.4
Picasso visiting Hofmann's exhibition at the Galerie Maeght, Paris, 1949.

Photo: E. L. White Jr.

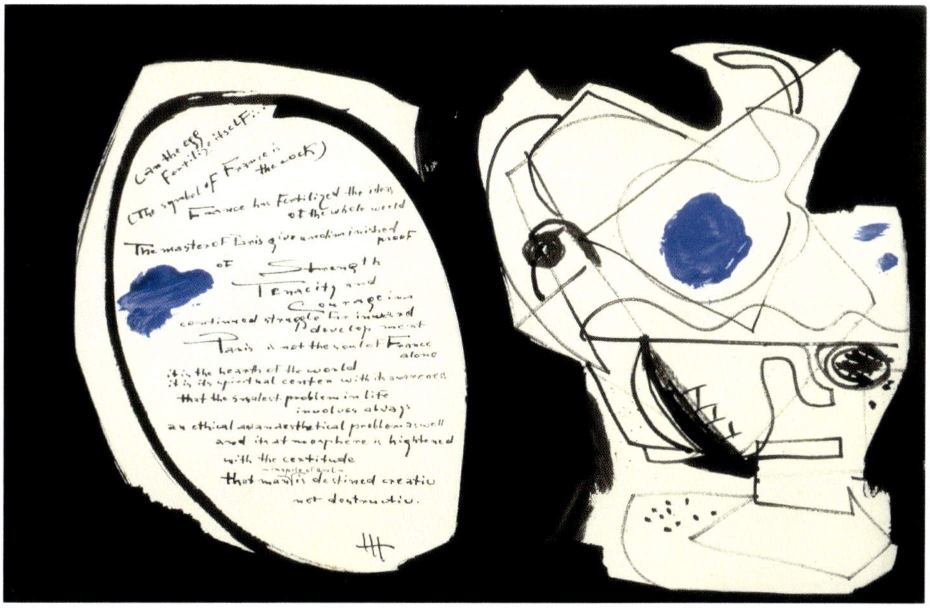

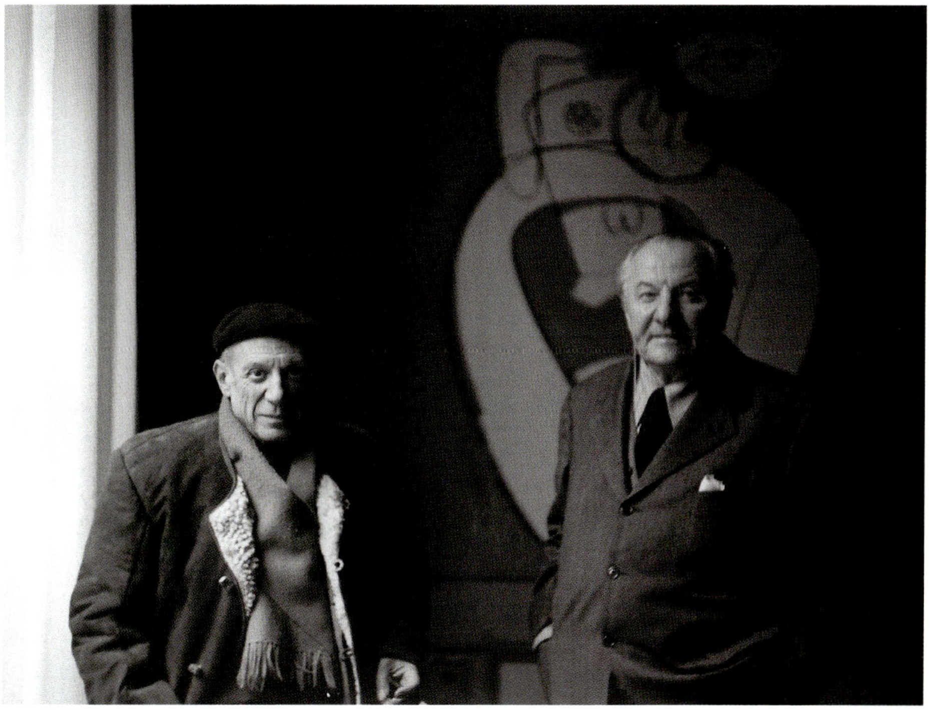

Terk and Delaunay were a link between French and German modernists. In 1912 the French poet Guillaume Apollinaire coined the term *Orphism* in reference to the explorations of pure color and abstraction practiced by Delaunay, Terk, and František Kupka.[14] "I will speak to you sometime about the subject of painting," Robert Delaunay wrote to Kandinsky as Sonia's translation of Kandinsky's influential *On the Spiritual in Art* was circulated in Paris, "[and] about an exciting conversation at the home of Apollinaire, who has begun to believe in us" (Dickerman 2012, 21). Like many artists of the period, Hofmann was influenced by Kandinsky's treatise, but few examples of Hofmann's work survive from this formative period in his career. Hofmann told Wight (1957, 30) that he spent the period working on still lifes, figures, and landscapes, and we can look to works

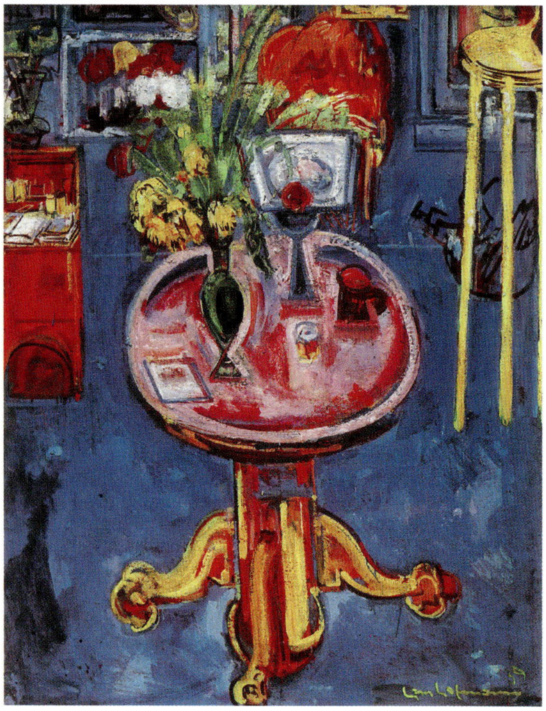

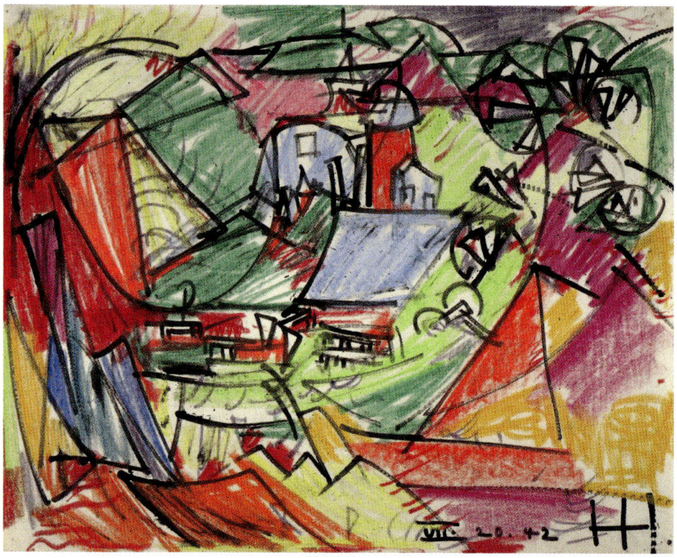

FIGURE 1.5
Hans Hofmann, *Interior No. 1—Pink Table, Yellow Tulips*, 1939. Oil on panel, 148.6 × 113 cm (58½ × 44½ in.). New York, collection of Mr. and Mrs. Preston Robert Tisch.

FIGURE 1.6
Hans Hofmann, *Untitled*, 1942. Crayon and ink on paper, 35.6 × 43.2 cm (14 × 17 in.). New York, Solomon R. Guggenheim Museum, 92.4043. Gift of Janet C. Hauck in memory of Alicia Guggenheim and Fred Hauck.

from Hofmann's early years in the United States (figs. 1.5 and 1.6) to see the influence of such Parisian artists as Cézanne and Matisse, as well as the Cubist composition, Fauvist color, and Post-Impressionist flattening of the picture plane that was to preoccupy Hofmann for the rest of his artistic career.

Return to Munich and a New Role Teaching Avant-Garde Art

The outbreak of the First World War in the summer of 1914 shattered cosmopolitan artistic communities in cities across Europe, including Paris. Hofmann returned to Munich as the advancing conflict scattered his French and German colleagues. He and Wolfegg were in Munich visiting his sister Rosa when war broke out and forced them to abandon their Paris studio. Upon their arrival, a friend of Wolfegg's, the artist Gabriele Münter, asked the Hofmanns to hold some of the artwork gathered up by Kandinsky before his departure, and in an ironic twist, while Hofmann lost his own paintings that were left behind in Paris, he was responsible for preserving many of Kandinsky's formative Munich creations (Dickey 2011, 52–53). "I would of course very gladly take advantage of your kind offer to choose a Kan[dinsky] for myself," Hofmann later wrote to Münter. "Up to now, the watercolor no. 5 has been hanging in our apartment-cum-studio.... I am especially fond of this picture" (Friedel 1997, 8–9). World War I also marked an end to Hofmann's support from Freudenberg, a change in situation that forced the artist to look for an alternative means of support. Excused from military service by a recent lung infection, Hofmann applied to the German government for permission to open an art school and was granted a government license on the condition that the school provide art therapy for shell-shocked veterans (Wessels 1967, 121).

Almost nothing is known of Hofmann's own painting during this period, as it seems he concentrated his energies on the needs of the new school. The Hans Hofmann Schule für Bildende Kunst (School of Fine Arts) opened in 1915 at 40 Georgenstrasse,

FIGURE 1.7
Hofmann (pointing at the easel) at the Hofmann Schule für Bildende Kunst (School of Fine Arts), Munich, March 1926. Worden Day Papers, 1940–82, Archives of American Art, Smithsonian Institution.

in the Schwabing neighborhood building where Hofmann himself had registered for Ažbe's classes years earlier (Goodman 1982, 27; Dickey 2011, 38). Hofmann's school filled the void left by the closure of schools run by foreign artists who were forced by the war to return home. As communication among artists broke down and art communities disbanded, Hofmann's firsthand experience with the formative art movements of both France and Germany quickly established his school as a direct connection to the lost zeitgeist of prewar avant-garde painting. "Students can probably learn more at the moment from Hofmann [in] Munich," American arts writer Sheldon Cheney would later note, "than [in] any school in Paris" (1924, 155).[15]

Enrollment, which was low during the war, improved in the 1920s with the relaxation of wartime travel restrictions and what educator Jennifer Cho called Hofmann's drive to "create his own student body" by recruiting from outside the traditional pool of local students (1993, 91). Hofmann's open enrollment policy offered student scholarships and welcomed older and female students, populations barred from attending the Munich Academy of Fine Arts (Singer 2003, 166) (fig. 1.7). The school's government license allowed enrolled students legal residency in Germany and swelled the school's postwar ranks with American students. Prominent American artists who traveled abroad to study with Hofmann in Munich included Louise Nevelson, Vaclav Vytlacil, scholarship student Alfred Jensen, and Wolfgang Paalen, who left his studies in Paris to go to Hofmann's Munich school (Goodman 1982, 29).[16] The school also recruited secondary and university art teachers for its summer sessions and soon established a reputation as a training ground for future educators. "Word had already gotten around," recalled American student Ludwig Sander (1969), "that if you studied with Hofmann you'd get a good job teaching somewhere."

The school curriculum—what Hofmann called "a re-won understanding of the principles of fine art...rooted in the work of our time" (Hofmann 1930b)—was built largely on the Cubist compositional theories and bold palette of Fauvist and Orphist colors that Hofmann had seen in Paris just before the war. Although brochures for Hofmann's Munich school advertise the program as "Expressionist" (Hofmann papers,

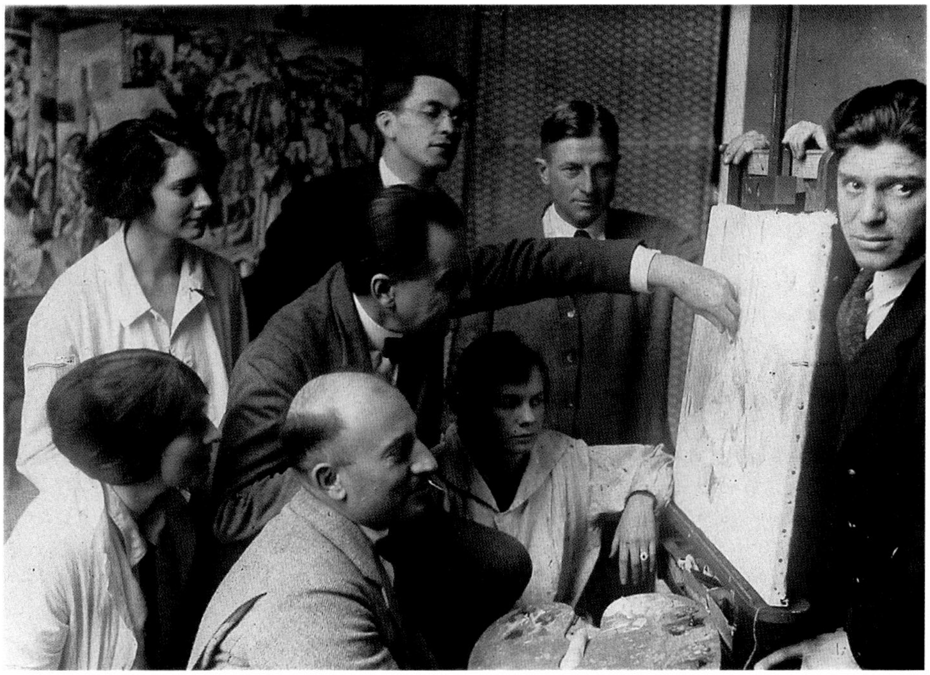

[ca. 1904]–78), common usage of that term was fluid. "The term suddenly appeared in relation to French painters and their exhibit in the Berlin Secession of 1911," recalled Selz, who confirmed that when Herwarth Walden invited "expressionist" artists to participate in a 1919 exhibition at his Der Sturm gallery, "the term still impl[ied] the contemporary French group, primarily Fauves" (1997, 256). "We call the art of this century Expressionism," wrote Walden, "in order to distinguish it from what is not art" (Selz 1997, 257). Hofmann's use of the term *Expressionist* in school advertisements was intended to align his school with French—or, more broadly, modern—aesthetics.

Munich's changing political climate endangered the livelihood of avant-garde art teachers, including Hofmann. The conservative nationalism that followed the war raised alarms within the avant-garde arts community throughout Germany. In the spring of 1919, members of Munich's artistic community, including Erich Mühsam, Gustav Landauer, and Ernst Toller—who was then the leader of the pacifist Cultural-Political Union of German Youth—took over the Bavarian Council's Republic and, according to Wessels, they "immediately... appointed Hofmann as the head of the Munich Academy." But the political impact of the modernists was short lived. "Two or three weeks later the Whites [the Soviet Räterepublik] took over again," Wessels recalled, "and they ousted [Hofmann], and he went back to his school in Georgenstrasse" (Wessels 1967, 122). Hofmann's popularity did not outweigh the dangers of his association with liberal activists and modern art. Hofmann was a leader of French modernism operating at a time in German society when much foreign and modern art was widely considered dangerous and subversive. In 1925, the Nazi Party set up offices at 50 Schellingstrasse, near Hofmann's school. At this time the artist's American students began an earnest campaign to bring Hofmann to the United States.

Journey to America

Hofmann made his first trip to the United States in 1930 to teach a summer session at the University of California at Berkeley, at the invitation of his student and art department faculty member Worth Ryder and with the assistance of fellow Hofmann alumni John Haley and Glenn Wessels.[17] Wessels acted as translator during the trip from New York to California and shared the driving duties with Hofmann. The drive was punctuated with lecture stops organized by Hofmann's students on art school faculties in cities including Chicago and Minneapolis.[18]

The Cubist painter and sculptor André Lhote had been scheduled to teach at the California School of Fine Arts that summer, and the cancellation of Lhote's classes made Hofmann the sole focus of students seeking a teacher with firsthand experience of Europe's formative modern art movements. More than two hundred students were turned away from Hofmann's oversubscribed classes at Berkeley, despite the extension of his summer session (*Art Digest* 1931, 29), and Hofmann was invited to return to Berkeley the following summer (fig. 1.8). Hofmann taught at Los Angeles's Chouinard Art Institute a full year before the arrival of the modern muralist David Alfaro Siqueiros, and as was the case with his Munich and Berkeley classes, Hofmann's Chouinard classes were heavily attended by both art students and art educators (Dickey 2011, 104) (fig. 1.9).

In various ways Hofmann was an early influence on modern art education in California. His first teaching manual was entitled Form und Farbe in der Gestaltung: Ein Lehrbuch für den Kunstunterricht (Creation in Form and Color: A Textbook for Instruction in Art). Hofmann began the text in Munich in 1915 and completed the manuscript in Berkeley in the summer of 1931. He then sent the manuscript to Berlin's Piper Publishing

FIGURE 1.8
Hofmann (pointing at the easel) during a Friday critique, University of California, Berkeley, 1931. Hofmann Papers, Bancroft Library, UC Berkeley.

Photo: Christina Lillian.

FIGURE 1.9
Hofmann (reclining at left) during summer session at the Chouinard Art Institute, San Pedro, California, 1932.

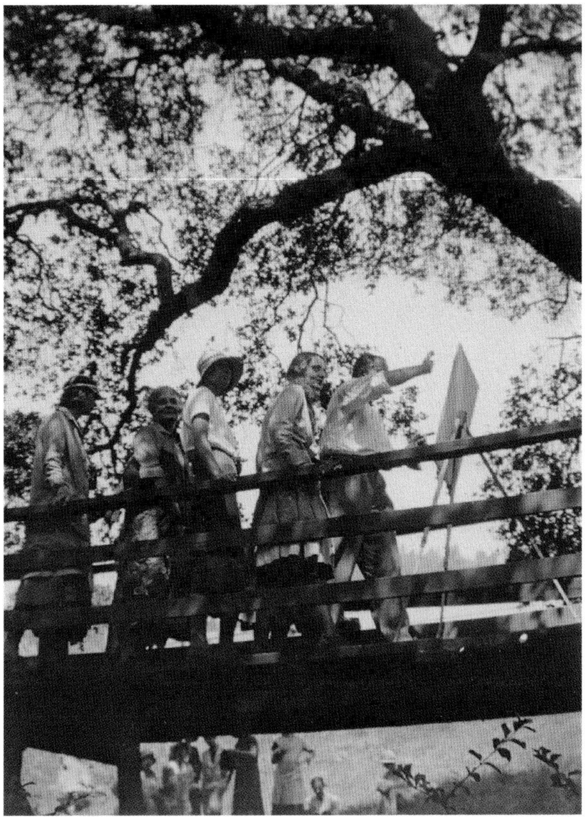

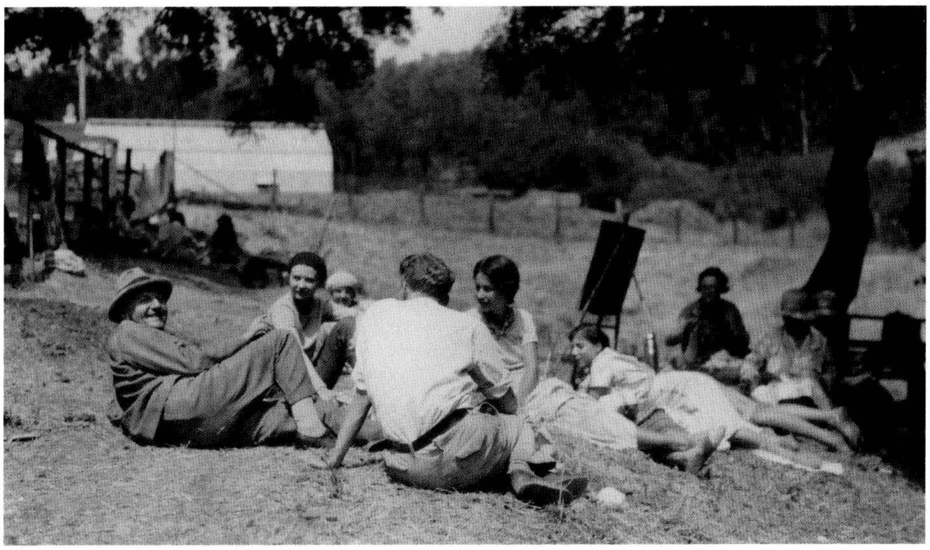

House (Goodman 1982, 66), which had published both the *Blaue Reiter Almanac* and *On the Spiritual in Art*. He gave copies to Wessels, who was then teaching at the California School of Arts and Crafts in San Francisco, and to Worth Ryder, who sought Hofmann's advice on department curricula (Wessels 1967, 290). Although Hofmann's manual was never published, sections were reproduced, in the West Coast arts publication *The Fortnightly*, in 1931 ("Painting and Culture") and 1932 ("On the Aims of Art"), and in the East Coast arts magazine *League*, in 1932–33 ("Plastic Creation") (Hofmann 1931, 1932, 1932–33). Hofmann's principles were also disseminated in publications by University of

California at Berkeley professor Erle Loran[19] and arts writer Sheldon Cheney, whose 1934 book *Expressionism in Art* (Cheney 1948) replaced Hildebrand's text as required reading for students in Hofmann's American classes. "[Hofmann's] ideas about space have been so widely disseminated by former students," Loran noted in a 1963 reprint of Cheney's text, "that anyone searching along similar lines necessarily owes a great deal to him" (Goodman 1982, 233).

Hofmann's impact on university curricula influenced the subsequent Berkeley School of Abstract Expressionism and may also have exerted some influence on the rival abstract painting scene in San Francisco, where California School of Fine Arts faculty included Hofmann's fellow Abstract Expressionists Mark Rothko, Ad Reinhardt, and Clyfford Still.[20] Still, for example, wrote his 1935 master's thesis on Cézanne's "synthetic development of color-form" using the principles outlined by Cheney the previous year—precepts that were directly influenced by Hofmann's teachings.[21] According to Wessels, "Still got some of his ideas from Cheney's book.... Thus Still is a 'student' by the extended influence of Hofmann" (Wessels 1967, 110). Hofmann's presence at Berkeley in the 1930s also served to establish a philosophical conduit between New York and the Bay Area more than a decade before the arrival of his colleagues in San Francisco, and years later Hofmann's influence continued to bridge the gap between opposing coastlines. "Hofmann taught at Berkeley and he taught all the teachers there, so in a sense Berkeley was an extension of the whole Hofmann... school," recalled land artist and Berkeley alumnus Walter De Maria, "so that we were very attuned to what New York's mind was like [when Hofmann was teaching there] in the Fifties" (De Maria 1972).

In the autumn of 1932 Hofmann traveled from California to the East Coast. He took a six-week teaching post at New York's Art Students League during the session when Jackson Pollock was elevated to the position of monitor in the mural class run by Thomas Hart Benton. By the time of Hofmann's arrival, the institution where such notable American artists as Benton and Stuart Davis taught had become a destination for students seeking exposure to European modernism. "We want foreign instructors," demanded students in letters to the editor of the League's paper during the 1932–33 winter session. "Few of us can afford to go abroad to study and learn what it is all about" (Dickey 2011, 105–6). Advance promotion by such former Hofmann students as League faculty member Vytlacil once again created for Hofmann an overflow audience, and extra class sessions were added to accommodate his growing roster of art students and members of the art elite, including Gertrude Henry, granddaughter of Gertrude Whitney, and Pegeen Guggenheim, daughter of Peggy Guggenheim. "Everybody studied with Hofmann," recalled Stuart Davis, who saw some of his own classes canceled due to low enrollment that season (McNeil 1965; Goodman 1982, 246).

In the fall of 1933 Hofmann opened a year-round school at 444 Madison Avenue; the Hans Hofmann School of Fine Arts spent a year at its inaugural location and brief periods at 137 East Fifty-Seventh Street and 52 West Ninth Street before settling at 52 West Eighth Street in 1937 for the remainder of its operation (Dickey 2011, 156–60). Hofmann's schools functioned as community centers for like-minded artists and provided students and teacher with ongoing exposure to the leading edge of art innovation. "It's not merely what you show," recalled Hofmann shortly before his death, "it's the process which is inherited in the created work that makes it a work of art" (Jaffe 1971, 35). The receptiveness to aesthetic and technical innovation that solidified Hofmann's reputation as "the dean of the abstract-expressionist movement" (*Time* 1956, 64; MoMA 1965, 6) was grounded in the artist's formative years in the avant-garde circles of Munich and Paris, and it would flourish in the American modernist community of mid-twentieth-century New York.

2 Hofmann and the Culture of Modern Painting

> Every intelligent painter carries the whole culture of modern painting in his head. It is his real subject, of which anything he paints is both an homage and a critique.
>
> — ROBERT MOTHERWELL, 1951

> There are values by which the new can be appreciated but they are not, in the first instance, esthetic values. Thus to appreciate the new *as the new* involves considerations beyond the esthetic.
>
> — HAROLD ROSENBERG, "The New as Value," 1963

Like many of his Abstract Expressionist colleagues, Hofmann was part of the twentieth century's "intellectual migration" of émigré artists and scholars whose flight from Europe to the United States in the years preceding the Second World War permanently raised the cultural profile of America in general and New York City in particular. In the late 1930s and early 1940s, the ranks of American modernists were swelled by refugee artists and art historians fleeing the war's advancing European front and facilitated by a network of earlier émigrés positioned at museums and academic institutions throughout the United States, with a particular concentration in New York City (Eisler 1969). As established American arts communities were joined by a wave of artists and scholars fleeing Europe, New York City emerged as the new global center of modernist experimentation and the center of the new direction in American art. When, in 1955, Princeton University student William C. Seitz wrote "Abstract-Expressionist Painting in America: An Interpretation Based on the Work and Thought of Six Key Figures"—a landmark dissertation on modern art focused on emerging styles of American painting—four of the six New York–based artists chosen by Seitz to represent "the direction taken in modern art in America since the Second World War" (Seitz 1955, i) were émigrés: the German-born Hofmann; Arshile Gorky, an Armenian; a Russian, Mark Rothko; and the Dutchman Willem de Kooning.[22] From the city's polyglot creative community emerged the groundbreaking collection of new ideas about art and art making that we call Abstract Expressionism, a group of innovative, seemingly disparate painting styles connected by what historian Timothy J. Clark called "a distinctive patterning of mental and technical possibilities" (1999, 7). Hofmann was older than many of the Abstract Expressionists—he was already an established artist and teacher when he moved to the United States[23]—and his years in Europe had provided him with an exposure to European modernism that went far beyond that of his American colleagues. "As time passes," reflected art historian and critic Barbara Rose (1978, 110), "the seminal role of Hans Hofmann in the development of Abstract Expressionism as an international style that fused elements derived from the major European modern movements becomes increasingly clear."

Hofmann's lifelong grounding in modernist methodology mobilized a generation of young avant-garde artists and positioned him at the center of America's emerging modern arts scene. "Through him more than any other one individual," writer and art critic Harold Rosenberg pronounced at Hofmann's funeral, "[this] was also a communion of artists" (1966, 21). Through his own efforts and the efforts of his students who were positioned at the forefront of art movements and institutions throughout the United States, Hofmann simultaneously influenced the leading edge of American modern art and served as its benefactor. "I think that the center [of post–World War II art] was influenced by the consciousness among artists in New York," former Hofmann student and New York Studio School founder Mercedes Matter noted, "and that consciousness had a great deal to do with Hofmann" (Dickey 2011, 322). Indeed, Hofmann would play a pivotal role in the creation of America's modernist community.

Abstract Expressionism is an umbrella designation commonly used to describe a range of innovative painting practices. Years before this American contribution to international modernism drew popular and critical attention, the term was applied to the shifting vanguard of German painting. It first appeared in a 1919 *Der Sturm* article about nature-based abstraction seen in the work of the German Expressionists, and within months the same phrase was used by German critics to describe the innovative Munich paintings of Russian émigré Wassily Kandinsky (Selz 1997, 342–43). In the United States, sporadic references to "abstract expressionist" German modernism began with a 1929 Wellesley College lecture by Museum of Modern Art founding director Alfred H. Barr Jr. (Selz 1997, 343), but the term failed to gain a foothold with American audiences.[24] The first use of *Abstract Expressionism* to describe American modernist painting appeared in a 1946 *New Yorker* review of a Hofmann exhibition at the Mortimer Brandt Gallery.[25] In his review, entitled "The Art Galleries: At Home and Abroad," critic Robert Coates proclaimed Hans Hofmann to be "one of the most uncompromising representatives of what some people call the spatter-and-daub school and I, more politely, have christened Abstract Expressionism" (1946, 83). Abstract Expressionism was thereafter associated with the mix of abstract composition and expressive paint handling that characterized mid-twentieth-century modern painting in America.

During the 1940s, avant-garde artists living in New York City transformed oil painting with experimentation in materials, technique, and composition. Hofmann drew far more than he painted during his early years in the United States, and while he resumed painting in the United States as early as 1934, his early work lacked focus.[26] The establishment and supervision of art schools in Munich and the United States imposed time and budget constraints that limited progress in his own artwork well into the 1940s. "The school as long as she not works as she should [*sic*] is always a great hindrance in the personal work," Hofmann confided in a July 14, 1941, letter to student Alice Hodges (Kiesler papers, [ca. 1910]–2003). Much of the artist's personal work during his early years in the United States consisted of landscape drawings and interior studies (see figs. 1.5 and 1.6). "For fifteen years [Hofmann] hardly picked up a brush," art critic Clement Greenberg noted, "but [he] drew obsessively" (1961, 12), and there are only a few existing examples of works on canvas prior to 1947, at which point canvas returned as a primary support material in Hofmann's work.[27] Hofmann sought value in this exploratory period. "The main thing is always to work and work," he wrote to Hodges and fellow student Lillian Olinsey Kiesler on September 5, 1941. "My work comes along in a rather experimental

FIGURE 2.1
Hans Hofmann, *The Wind*, 1937–42. Oil, Duco, gouache, and India ink on posterboard, 111.4 × 70.5 cm (43⅞ × 27¾ in.). Berkeley, California, Berkeley Art Museum and Pacific Film Archive (BAM/PFA), 1965.16. Gift of Hans Hofmann.

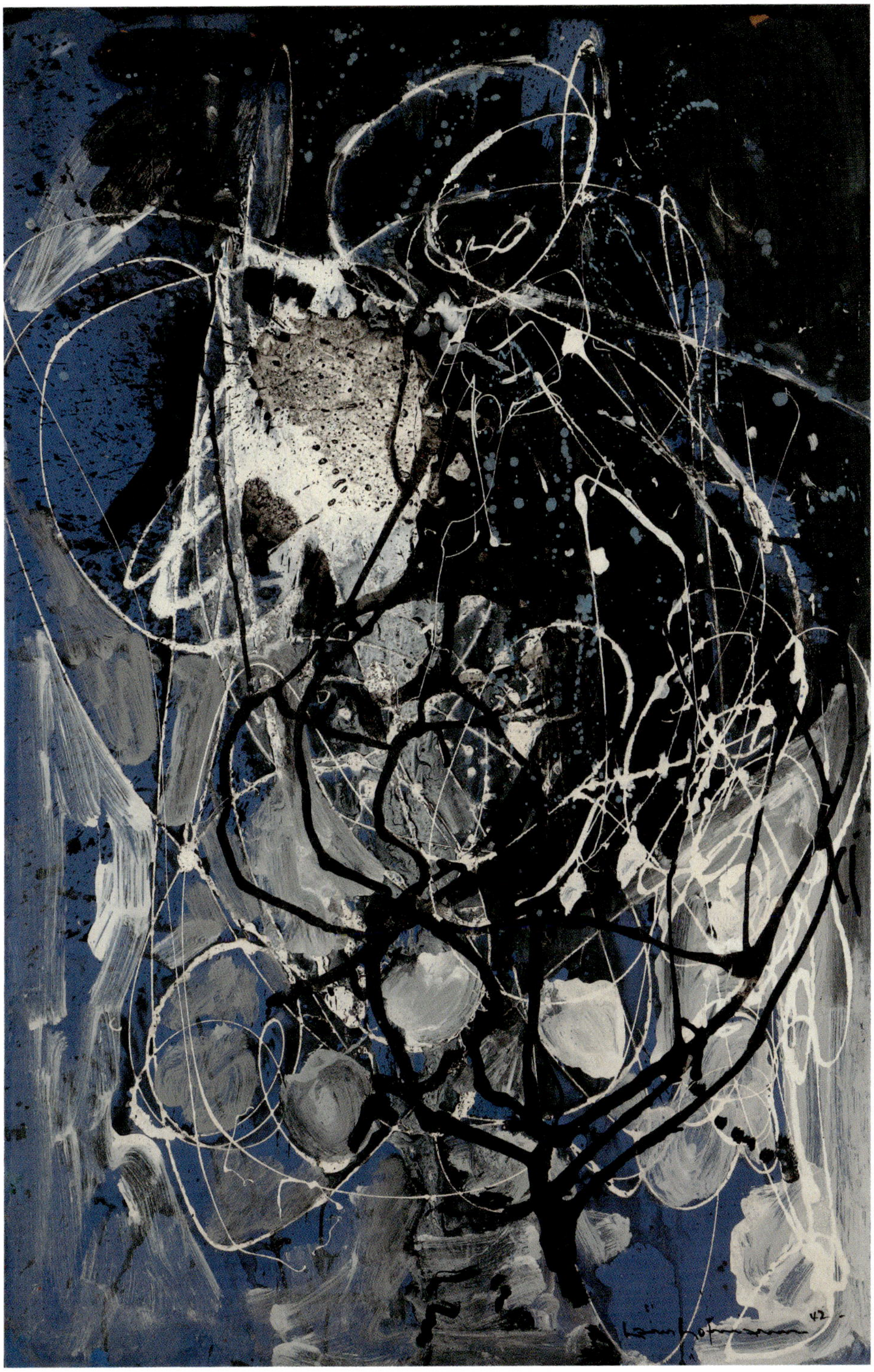

period [in] which I find myself [placed] on the way to the highest freedom" (Hofmann papers, [ca. 1904]–78). The range of styles explored by Hofmann upon his arrival in New York have placed the artist among every midcentury modernist movement outlined in Irving Sandler's compendium *The Triumph of American Painting: A History of Abstract Expressionism* (Sandler 1970). Hofmann was not alone in this search; his automatist experiments from this period, for example, bear a striking resemblance to works by such

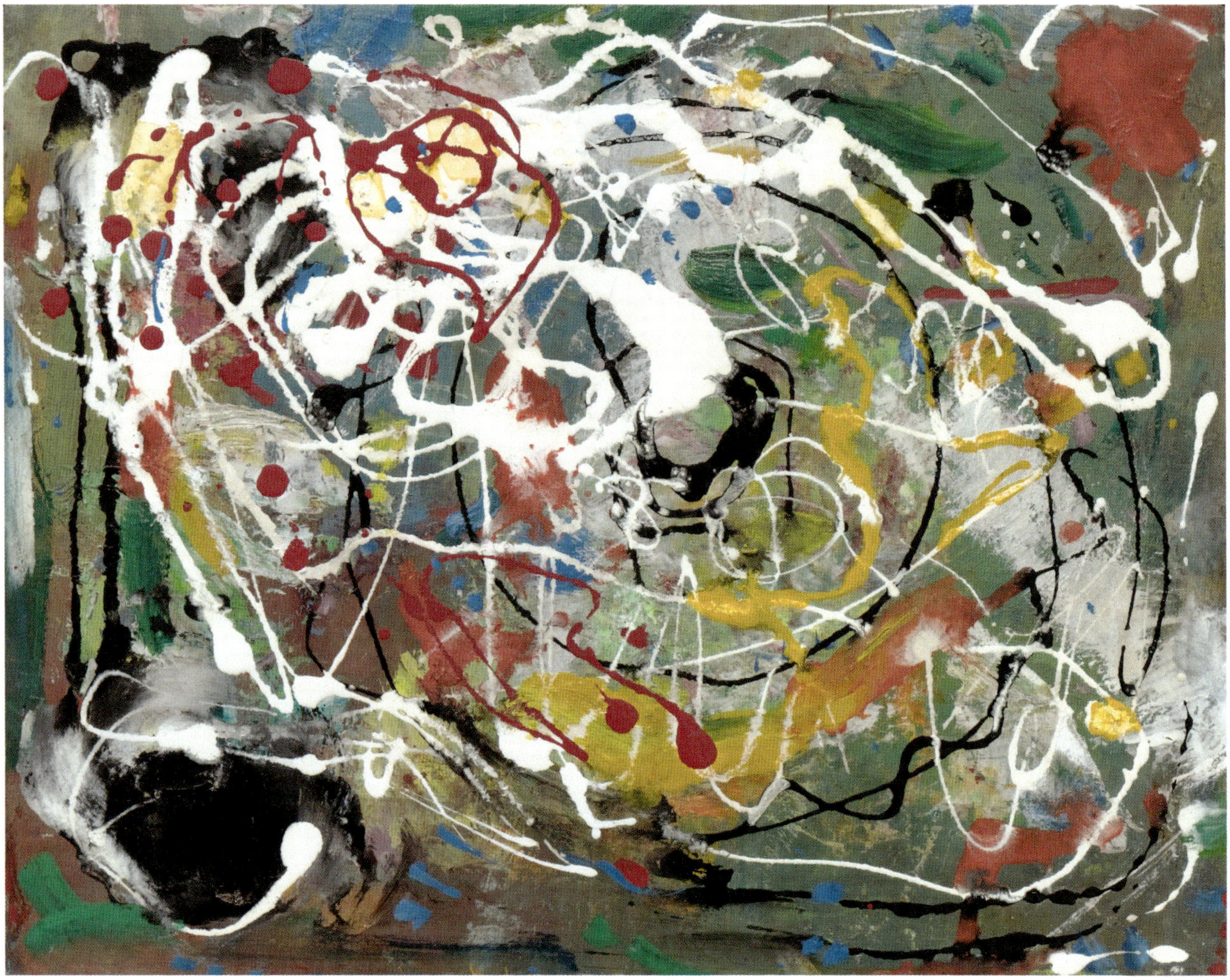

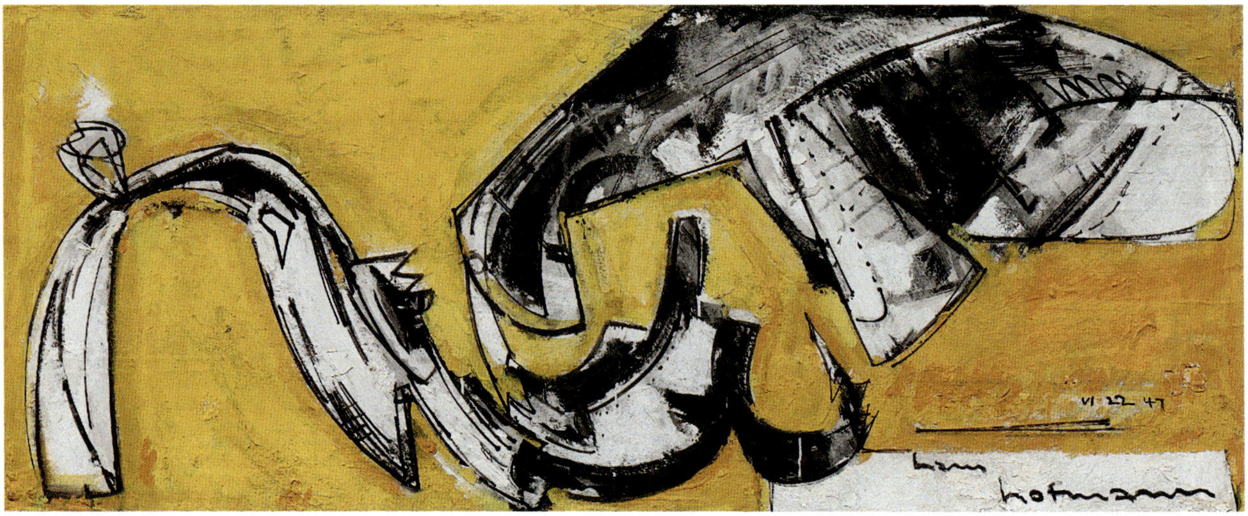
VI·22·47
hans hofmann

FIGURE 2.2
Hans Hofmann, *Spring*, [1940] 1944–45. Oil on panel, 28.6 × 35.9 cm (11¼ × 14⅛ in.). New York, Museum of Modern Art, 1516.1968. Gift of Mr. and Mrs. Peter A Rübel.

FIGURE 2.3
Hans Hofmann, *Embrace*, 1947. Casein on panel, 60 × 144.8 cm (23⅝ × 57 in.). Boston, Museum of Fine Arts, 1990.416. Gift of the William H. Lane Foundation.

New York–based artists as Jackson Pollock, Janet Sobel, and André Masson (figs. 2.1 and 2.2), just as his contemporaneous biomorphic works show similarities with the ambiguous organic forms found in paintings from the same period by colleagues including Robert Motherwell, William Baziotes, Gorky, and de Kooning (fig. 2.3).

The range of experimentation prevalent among nascent Abstract Expressionists was noted by William C. Seitz during his tenure as a curator at the Museum of Modern Art, when he wrote a catalogue essay for the museum's Hofmann retrospective.[28] "When one looks back at the years after 1945, when the 'New American Painting' was taking form," Seitz noted, "it is apparent that one of its aesthetic determinants was the desire felt by many artists to incorporate in their work tendencies of style and feeling previously thought to be contradictory" (1963, 7). This unconventional environment was similar to that of Hofmann's years in Paris, which he described as "a wonderful confusion" (Dickey 1997, 85).

Sandler noted that similarities in early experimentation—or "multiple discovery"—shared among the New York artists created difficulties for critics:

> The most difficult "fact" to determine is who painted what first. It is generally assumed that if artist A worked prior to artist B in a certain style, then artist A influenced artist B, and that A is a leader and B a follower. This is frequently true, but it also may happen that artists A and B were friends who exchanged ideas, or that artist B was tending in a direction similar to artist A before seeing his pictures, or that artist B, responding to the same stimuli in art and in the world at large, developed independently of artist A. (1970, 3)

Sandler and others noted that efforts to cite ownership of such techniques avoid more fruitful discourse. The drip technique associated with its masterful use by Pollock, for example, can be traced to numerous contemporary practitioners, including Hofmann and Gorky (Hunter 1959, 136), Mexican muralist David Alfaro Siqueiros (Sandler 1970, 113; Buettner 1981, 96), Surrealist painter Max Ernst (Varnedoe 1998, 49), and Janet Sobel, whose work Greenberg claimed Pollock admired "rather furtively" (Sandler 1970, 113). While arguments pitting Hofmann against Pollock for primacy in this matter continue,[29] critics and curators including Sandler and Hunter were steadfast in their belief that "it is a reproof to neither artist" to recognize the communal search from which individual Abstract Expressionist artists found a personal style containing the "psychological intensity and . . . individual expressive content" to make a lasting contribution to art history (Hunter 1964, 17).

The term Abstract Expressionism was embraced by the press and the general public despite the artists' objections. "This seems to be the old situation," Harold Rosenberg remarked at a meeting at The Club, "where the critic tries to round up the painters [into categories] and the painter tries to get out" (Seitz 1955, 404). The artists deemphasized the commonalities in their work and worried aloud that a single descriptor would stifle individual expression. "It is disastrous to name ourselves," de Kooning told a group of prominent modernists assembled in April 1950 for the now-famous Artists' Sessions at Studio 35, closed-door meetings of artists and critics centered around discussions of Abstract Expressionist work and community (fig. 2.4). "Everyone should be as different as possible," agreed Hofmann. "There is nothing that is common to all of us except our creative urge" (Motherwell, Reinhardt, and Karpel 1951, 22, 10). Throughout the present volume, the work of Hofmann and his contemporaries is designated Abstract Expressionist, not only because of the term's familiarity but also because the composite name highlights the movement's primary areas of innovation—composition and technique—and

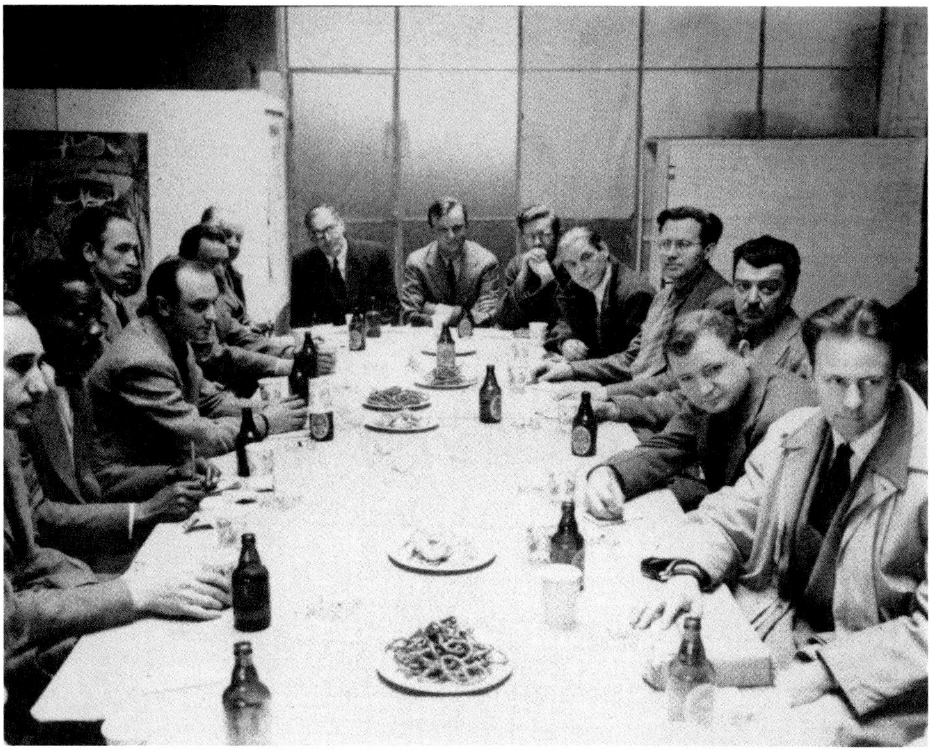

FIGURE 2.4
Artists' Sessions at Studio 35, New York, April 21–23, 1950. From left to right: Seymour Lipton, Norman Lewis, Jimmy Ernst, Peter Grippe, Adolph Gottlieb, Hans Hofmann, Alfred H. Barr Jr., Robert Motherwell, Richard Lippold, Willem de Kooning, Ibram Lassaw, James Brooks, Ad Reinhardt, and Richard Pousette-Dart.

Photo: Aaron Siskind.

emphasizes the communal nature of modernism more successfully than alternatives that promote the movement's isolating aspects, including *New York School*, *action painting*, and *American-type painting*. Despite de Kooning's prediction of disaster, the name Abstract Expressionism has served to bring public attention to a range of experimental artworks and fix the related achievements of myriad artists within the flow of history.

Hofmann's arrival in New York again placed him both temporally and physically at the center of modern art practice. During the 1940s and '50s, his painting studio was located near fellow artists Franz Kline and Ad Reinhardt and modern art critics Rosenberg and Dore Ashton, and close to his own school as well as to the Subjects of the Artist School—a presentation forum established just a block away by Baziotes, Motherwell, Rothko, Clyfford Still, and David Hare, and which had a subsequent incarnation as The Club.[30] From 1941 through 1945, Hofmann's studio was located downtown at 44 East Eighth Street (Dickey 2011, 184), adjacent (although located in the building next door) to that of his student Lee Krasner and her partner Pollock,[31] through whom Hofmann met younger artists, including Motherwell, who described their first encounter:

> Pollock was living with Lee Krasner, who was a pupil and admirer of Hofmann, who lived only a few doors from them. I think it was at Lee's behest that Pollock took me over after dinner to see Hofmann. It was the first time I had met him, and I at once realized that my being 26 years old—Hofmann was already in his 60's—it would be impertinent for a young apprentice artist to tell him about what painting was. As it turned out, Pollock got drunk on a big jug of red wine, and we all had to carry him down four flights into the street and then up four flights to his place. It was a helluva job. (Simon 1967, 22)

Hofmann's Eighth Street neighborhood was a social and professional hub that housed many of the artists' studios, private classrooms, and public schools. The artists met for coffee at the Waldorf Cafeteria and for drinks at the Cedar Tavern, and they dropped in on Friday critiques and lectures a few doors down at Hofmann's school. Friday lectures by fellow artists were also held at the Subjects of the Artist School, and when administration of that space was handed over to faculty from New York University the following year, less formal meetings began two doors down at an address referred to as the Eighth Street Club, the Artists' Club, or simply "The Club."[32]

Assisted by Hofmann's 1935 opening of a summer school on the grounds of the old Charles Hawthorne artists' colony in Provincetown, Massachusetts, the Abstract Expressionists were an integral part of the rebirth of that summer destination for the New York City art world. Provincetown summer residents included Hofmann, Kline, Pollock, Krasner, Motherwell, and Helen Frankenthaler. The Cape Cod arts community, following in the footsteps of the Subjects of the Artist School and The Club, staged notable events including Forum '49, the 1949 summer-long program of exhibitions and debates focused on progressive American art.

Hofmann's Role in Abstract Expressionism and His Influence on Generations of Artists

Hofmann did not share the social activism of such colleagues as Reinhardt and Barnett Newman, but he was an active participant in the modernist cause. He was part of the renowned Studio 35 roundtable sessions in 1950 and one of the so-called Irascibles, the group of eighteen painters and ten sculptors who in the same year penned an infamous public letter to Metropolitan Museum of Art president Roland L. Redmond protesting the museum's "hostility to advanced art."[33] During his lifetime alone, Hofmann's work was exhibited in fifty solo shows in the United States and Europe, as well as in countless group shows, including most of the formative exhibitions of the burgeoning Abstract Expressionist movement. Hofmann was an early member of the Betty Parsons Gallery and was later part of the stable of artists promoted by Samuel M. Kootz, who was the American representative for Pablo Picasso.[34] Hofmann's first solo painting exhibition in New York took place in March 1944 at Peggy Guggenheim's Art of This Century gallery.[35] It was one of the early exhibitions that Greenberg later said had signaled the future of "American-type painting" (1955). Hofmann's work was included in Sidney Janis's 1944 touring show *Abstract and Surrealist Art in America* and Howard Putzel's showcase in the same year of emerging and early American modernist painting entitled *Forty American Moderns*. Works by Hofmann were also part of Howard Putzel's 1945 proto–Abstract Expressionist exhibition *A Problem for Critics* and in a similar showcase curated by Newman at the Betty Parsons Gallery entitled *The Ideographic Picture*—exhibitions notable for their focus on the new direction of modern art.[36] "Spontaneous and emerging from several points, there has arisen during the war years a new force in American painting," proclaimed Newman in the exhibition announcement. "It is now time for the artist himself... to make clear the community of intent that motivates him and his colleagues" (1947).

Critical reception of Hofmann's work initially suffered from the artist's use of small canvases. Hofmann continued to work with conventionally sized canvases—those that could fit on an artist's easel and hang on a modestly sized wall—long after his Abstract Expressionist colleagues had shifted to the large-scale compositions associated with an artist's direct engagement with painting. "To paint a small picture," Rothko told an audience at a Museum of Modern Art symposium,[37] "is to place yourself outside of your experience" (Rothko 1951, 104). Hofmann's lingering loyalty to what Greenberg called "the easel convention" (1955, 184) was seen as an impediment to Hofmann's direct engagement with nature, the ultimate source of abstract subject matter. "You see, Hofmann separates himself from nature," Lee Krasner recalled, "he puts nature out there, and he is the observer" (Diamonstein 1979, 202).[38] The rising prominence of a bold palette and a distinctive compositional vocabulary in Hofmann's paintings, however, soon quelled concerns regarding the size of the artist's work. "[Hofmann's] canvas, in point of fact, is man-sized," Seitz noted in his 1955 dissertation, "[but] there is no doubt that...his best canvases scale larger" (1955, 233, 240). Hofmann's vivid palette and bold compositions soon set his works apart from those of his Abstract Expressionist colleagues. Although his canvases never grew to the size of those by Pollock or Rothko, the distilled color and form of Hofmann's larger, late paintings often overpowered nearby works by other artists. "At more than one group show," recalled Greenberg (1961, 28), "I have had the experience of seeing even a rather indifferent Hofmann make all the other works present, including those by more cried-up artists, seem a little less than present by contrast." Hofmann's late-career paintings are a combination of both medium and message—fundamental truths explored through the fundamental materials of painting. "When I myself not so long ago complained in print that Hofmann was failing to realize his true potentialities," Greenberg admitted, "it was because I had not caught up with him" (1955, 184).

A primary characteristic of mid-twentieth-century art is the evolving relationship between artist and material, and any thorough investigation of Hofmann's paintings must consider his exposure to new materials and the avenues of informational transfer regarding those materials. The arts community that grew around Hofmann's American schools placed him at the center of contemporary art theory and practice. Rosenberg stated that any study of the arts "must maintain a continuing sensitivity to major characteristics peculiar to the modern epoch that affect the situation of art" (1975, 135). Hofmann's formative years in the burgeoning avant-garde scenes of Europe brought him into contact with groundbreaking advances in modern art and science and gave rise to his lifelong support of emerging artists and the communities that facilitated their exposure to innovative art-making materials, techniques, and theory.

Hofmann taught continuously from 1915 to 1958 but is best known for energizing American artists in the 1940s and 1950s with lectures and theories that drew on his firsthand experience with European modernism. "[Hofmann] interpreted the ideas of Fauvism and Cubism, as well as those of German Expressionism and Surrealism, in terms of...the radical post–World War II approach to painting," claimed Sandler. "In fact, Hofmann was in large measure its originator" (1963, 15). Hofmann's schools in New York and Provincetown drew artists and critics from around the country to study with the renowned teacher and learn his method of adjacent color and form placement to achieve the appearance of dimensionality, a methodology better known as "push and pull." Hofmann quickly became the "foremost figure in the education of modern artists in America," noted Seitz, and his schools emerged as "primary centers for the study of modern painting" (1955, 12). In the words of Lillian Olinsey Kiesler, who studied with Hofmann from 1932 to 1947, "I was walking on Eighth Street and along came one of the people from the [Art Students League] class who said 'The greatest art teacher in the world is coming.'...You can be sure the first day that Hans Hofmann appeared...I was there together with sixty other people" (Bultman 1978).

Hofmann's New York school was "by all accounts a magnet that drew visitors constantly," according to Ashton. As mentioned above, Friday evening critiques at the school—situated between The Club and the Cedar Street Tavern—were a popular destination for artists and critics. "Many of the evening sessions during which Hofmann offered impromptu lectures," recalled Ashton, "were attended by the artists in the neighborhood,

FIGURE 2.5
Crowds attending a public Friday critique held on the patio of the Nickerson Street studio, Provincetown, Massachusetts, ca. 1950.

Photo: Ellen Buchman.

FIGURE 2.6
Hofmann responding to a painting during a public Friday critique at the Nickerson Street studio, Provincetown, ca. 1950.

Photo: Arnold Newman.

who would then move on to the local cafeteria to continue the discussion" (1972, 79). Hofmann's school was also near the progressive collections of A.E. Gallatin and the contemporary American art exhibitions of the Whitney Museum of American Art (formerly the Whitney Studio), and the school's convenient downtown location brought frequent visitors who kept Hofmann and his students abreast of current trends and community interests.[39] "Around the corner was Hans Hofmann's school," recalled sculptor Philip Pavia. "Everybody went to his school and picked up the language of Hofmann. With that started this nucleus" (de Antonio and Tuchman 1984, 39).

Hofmann opened his Provincetown location in the former Hawthorne studios in 1935 and fashioned a summer arts center, discussed above, which educated "not only a generation of artists," according to *New York Times* critic Hilton Kramer, "but also a number of critics, museum directors, and collectors too. His school was a cultural force." "Mr. Hofmann was, in effect," said Kramer, "one of the forces that completed America's education in modern art" (1966, 33). Capacity audiences at Hofmann's lectures and public critiques spilled into the street and drew regular crowds of onlookers (Goodman 1986, 57; Dickey 2011, 307) (figs. 2.5 and 2.6), and his students reported that when Hofmann participated in the Provincetown school's opening panel for "Forum '49"—entitled "What Is an Artist?"—five hundred people were turned away for lack of space (Seckler 1977, 64–65). The fervent appetite for Hofmann's pedagogy was described by James Gahagan, who studied with Hofmann from 1952 to 1958: "I stopped off in … Albuquerque to visit a couple of friends in the spring and they en masse were loading themselves into two automobiles to come back into Provincetown and study with this man, Hans Hofmann.… I jumped in the car with them and came right back again" (Bultman 1978).

Unofficial tallies of school rosters number Hofmann alumni in the thousands. Hofmann's former students include more than half of the founding members of the American Abstract Artists group,[40] including Josef Albers,[41] Ray Eames, and New York Studio School founder Mercedes Matter.[42] "Half of the avant-garde in the 1930s was enrolled in [Hofmann's] classes," noted Sandler (2009, 56). Students at the Hofmann school were also founding members of the Flatiron district and Tenth Street cooperative galleries that promoted the emerging Pop and environmental art movements.[43] Students include artists of note in a wide range of experimental media, including installation art

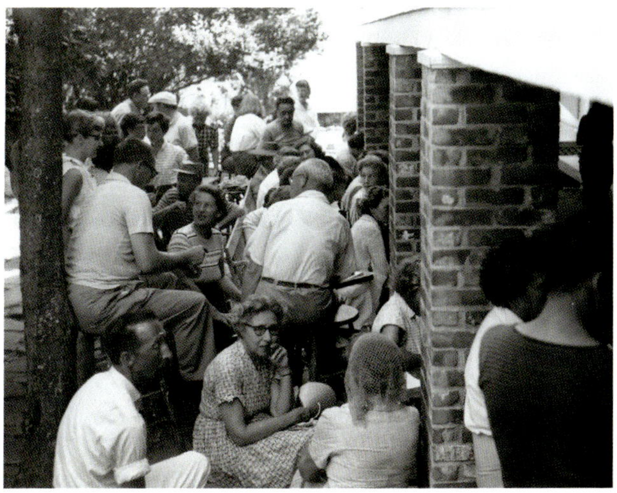

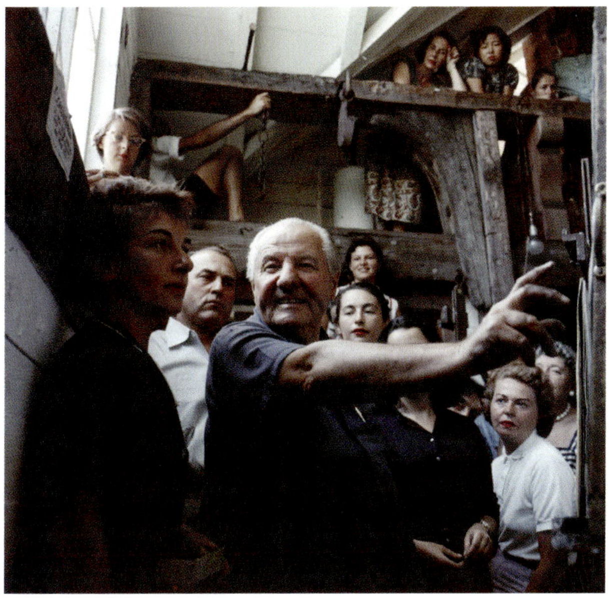

Modern Paint Technology in the United States

America's mid-twentieth-century artists benefited from both advancing paint technology and a concentrated effort by American manufacturers to promote new materials directly to artists. Wartime restrictions on the phenol- and alkyd-based paints embraced by such French modernists as Pablo Picasso limited, but did not completely prevent, the access of Abstract Expressionist painters to experimental formulations. In 1936, for example, Jackson Pollock and Morris Louis worked with synthetic paint binders at the Laboratory for Experimental Techniques in Art, held by David Alfaro Siqueiros at 5 West Fourteenth Street (Marontate 1996, 60); this was a forum in which artists could speak with paint chemists and work with the ethyl silicate paint produced by Keim Mineral Coatings[f] as well as the new pyroxylin-based paints of military coating manufacturer E. I. du Pont de Nemours and Company (later DuPont).[g]

Shortly thereafter, researchers for the Works Progress Administration (WPA) developed their Glyptal mural paint based on the alkyd house paint of the General Electric Company (now GE), whose patent on the formulation had run out in 1935 (Standeven 2007, 78). The origin of Glyptal paint is alternatively attributed to teacher Frank Sterner, director of the WPA's Boston Paint Testing and Research Laboratory (Lodge 1988, 121), and to Raphael Doktor, head of the WPA's Restoration, Installation and Technical Service Division in New York City (Marontate 1996, 94). Artists in the Painting and Sculpture portion of the government program—including William Baziotes, Willem de Kooning, Arshile Gorky, Adolph Gottlieb, Philip Guston, Lee Krasner, Pollock, Ad Reinhardt, and Mark Rothko—would likely have worked with Glyptal and similar commercial resin paints in the execution of WPA projects, and some artists were still using the material years later in their own work. "His medium is a mixture of one part of stand oil to two parts of damar varnish and two parts of oil of gum turpentine," recounted Fairfield Porter of Jack Tworkov's technique in the 1953 *ARTnews* article "Tworkov Paints a Picture" (1953, 73): "[Tworkov] dissolves areas to be removed with Glyptal thinner, and adds Glyptal in small quantities to the paint to make it dry faster."[h]

Commercial materials were in common use among the Abstract Expressionists, many of whom had worked in commercial trades. De Kooning, for example, supported his early years in New York with work as a sign painter, while Rothko's training in scenery production was adapted to his signature painting technique, which, according to Thomas Hess, involved the staining of canvas with "scenery paints cooked over a hot-plate" (Ashton 1995, 43; Hess 1970, 29). Pollock[i] and Franz Kline (Goodnough 1952, 38–39) became associated with preferred commercial paint brands despite the efforts of such commercial gallerists as Sidney Janis, who, according to Irving Sandler (2012), once broke into Kline's studio and replaced the artist's house paints with tubes of traditional Winsor and Newton oil colors. Press coverage at the height of the movement's popularity carried news of the popular new paint materials to audiences around the globe (Standeven 2003, 58).

The growing commercial paint market was not lost on postwar coatings manufacturers, who found themselves with idle factories and a surplus of materials. The range of art materials available to New York's modernist painters expanded dramatically after 1945, with both an increased number of commercial coatings manufacturers and the introduction of small-run and custom paints marketed directly to artists by local paint makers including Leonard Bocour and Sam Golden of Bocour Artist Colors and Henry Levison of Permanent Pigments (Marontate 1996; Stoner 2004). These small-run paint makers played a pivotal role in the development of artists' acrylic paints and their use by Abstract Expressionists and other members of the New York arts community. The use of poly(vinyl-acetate)-based paints by New York artists was noted in a 1939 article in *The*

Art News (later *ARTnews*) (Lodge 1988, 124), but their popularity was limited, and artists' use of vinyl emulsion paints waned with the postwar introduction of acrylic paint media. As with vinyl resin paints, acrylic resin paints were first available only in resin-in-solvent formulations, followed later by the commercial distribution of waterborne acrylic emulsion paints.

In 1947 Bocour and Golden released Magna, an acrylic-resin-in-solvent paint marketed as "the first new painting medium in 500 years"; it was anecdotally based on a sample of Acryloid F-10 resin, then in use on WPA projects.[j] The new paint was included in Bocour's lectures around the country and in "Bocour Workshop" classes at the Brooklyn Museum Art School and the Skowhegan School of Painting and Sculpture, and samples of paint were given to artists who tested the material. Notable Abstract Expressionist users of Magna included Barnett Newman, Rothko, and Robert Motherwell, as well as Morris Louis and fellow Washingtonian Kenneth Noland.

In 1954 Permanent Pigments began commercial production of Liquitex—an artist's emulsion paint based on Rohm and Haas AC-33 acrylic resin (Lodge 1988, 124)—which contributed to the change in Helen Frankenthaler's style when she switched from oil to acrylic in 1963. "When first introduced, Liquitex was rather thin and runny, and was not immediately successful," according to conservators Jo Crook and Thomas Learner. "However, several artists did experiment with it, such as Andy Warhol and Helen Frankenthaler. . . . Frankenthaler's second trial with acrylic emulsion paint in the early 1960s [Liquitex introduced a new formulation in 1963] was successful and from then on she continued to paint with it" (Crook and Learner 2000, 29).

Paint makers were an active part of the New York arts community; they made presentations, taught classes, hired artists, and visited artists' studios. As Bocour recalled (1985), "Everybody who was anybody came, passed through [our] doors."

FIGURE 2.7
Hofmann giving a demonstration at an easel, Nickerson Street studio, Provincetown, ca. 1950.

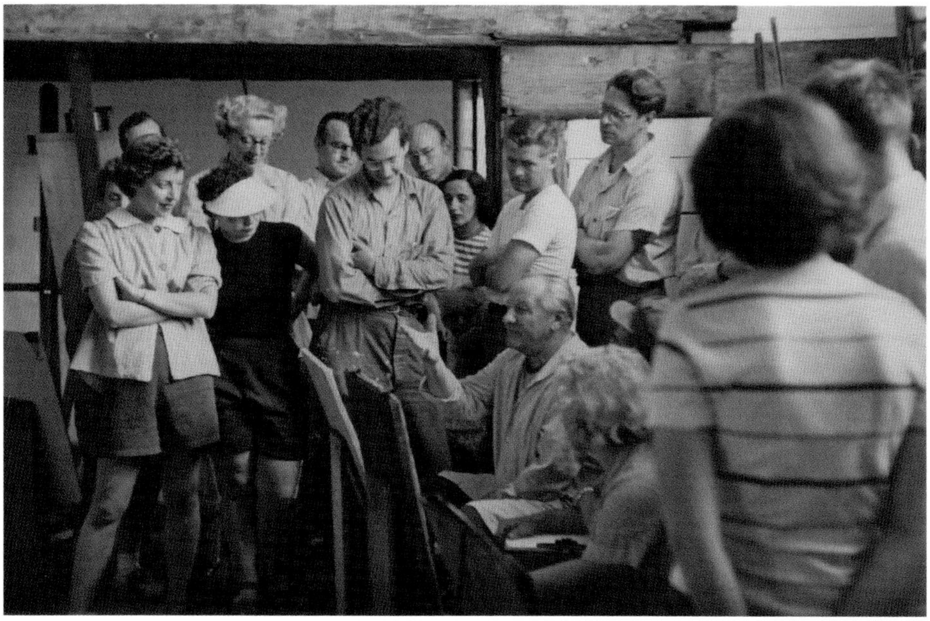

pioneer Red Grooms, Happenings creator Allan Kaprow, and Minimalist Daniel Flavin. "If a teacher's stature is measured by the number of students who achieve national and international renown in their own right," proclaimed Sandler (1973, 49), "then that of Hofmann is without equal."

Hofmann not only directly influenced thousands of students but also indirectly affected countless others through the teachings of former students at the forefront of art movements and institutions throughout the United States. He "influenced the curriculum [of] generations of students," said land artist Walter De Maria. "He taught all the teachers…[who then] could go out and teach it to a thousand other people" (De Maria 1972) (fig. 2.7). "I was a Hofmann student without knowing I was one," recalled Minimalist painter Frank Stella (Amgott 2003), who studied at Princeton under the mentorship of Hofmann scholar Seitz. At the age of seventy-eight and after forty-three consecutive years of teaching, Hofmann closed his schools in 1958 and began to paint full-time, but the ties that bound Hofmann to generations of modern artists remain intact. "We can derive novel conclusions," Rosenberg posited shortly before Hofmann retired, "from reflections of [Hofmann] sent back to him by artists whose inventions have been in debt to his techniques" (1957, 56).

The Role of Materials in Abstract Expressionism

Abstract Expressionist paintings share two primary characteristics: abstract compositional elements and expressive painting technique. Seeing these traits in the work of the disparate Abstract Expressionist artists is central to understanding the formative role the movement played in modern art. Abstract Expressionist painting is at its root representational. Even the most elemental work of such American Abstract Expressionists as Rothko and Newman is primarily abstract—not nonobjective—and despite confusion in the popular press and even in some current-day scholarly publications, the terms are not interchangeable. An early description of the differences between these terms appears in the 1943 book *New Frontiers in American Painting* by Samuel M. Kootz, whose eponymous gallery featured the work of such prominent Abstract Expressionists as Hofmann,

Motherwell, and Baziotes. Kootz described abstract painters as using "existent reality as a point of departure for their abstractions, and insist[ing] upon a close attachment to life-impulses," while nonobjective artists "attempt to find perfection in geometry alone, with no recognition of humanity" (1943, 49–50).

Kootz was Picasso's American representative, and he used the proceeds from that lucrative arrangement to subsidize the early work of such Abstract Expressionists as Motherwell and Baziotes (Kootz 1964). Perhaps Kootz's clarity of thought was based on insights provided by his renowned client. "You must always start with something," Picasso remarked in 1935; "afterwards you can remove all appearance of reality; there is no longer any danger, because the idea of the object left an indelible mark" (Bois 2012, 40). Even the name of the Abstract Expressionists' discussion forum—"The Subjects of the Artist"—was a nod to that commitment. "The title was Barney's [Newman's]," recalled Motherwell (Carmean 1978, 15), "[and] we all agreed that it was right because it made the point that our works did have subjects."

As their early experiments with subject-based ideographic, biomorphic, and mythic compositions gave way to recognizable signature styles, the mature Abstract Expressionists may at first glance appear to abandon content for construction. However other-worldly Still's topographic paint mounds, Rothko's ethereal glazes, or Newman's monochrome expanses may have appeared to contemporary audiences, these works remained rooted in the exploration of nature and the human experience. "There is no such thing as a good painting about nothing," claimed Rothko (Jewell 1943, x9). Through individualistic brushstrokes, paint splashes, and pigment stains, the Abstract Expressionists strove together to distill and convey the quality of human experience. "The creative process," noted Hofmann (1931, 5), "lies not in imitating, but in paralleling nature—translating the impulse received from nature into the medium of expression." While some Abstract Expressionists returned to more overt subject matter late in their careers, at the movement's peak the painting of Hofmann and his colleagues expanded the boundaries of acceptable subject matter by creating a gateway between literal and experiential representation in modern art.

The innovative application of paint is a recognizable characteristic of Abstract Expressionism. Unconventional painting tools and media were adopted by New York artists during this period, including experimental paint formulations and industrial materials previously considered outside the realm of fine arts practice. De Kooning's painting technique, for example, incorporated the sign-painting tools of his early work as a commercial artist, while the experimental paint workshops held by Siqueiros in New York played a formative role in Pollock's embrace of nontraditional paint media. The many Abstract Expressionist painters employed through New Deal programs at the Treasury Department or the Works Progress Administration (WPA)—including Baziotes, Gorky, Adolph Gottlieb, Philip Guston, Kline, Krasner, Pollock, Reinhardt, Rothko, and Mark Tobey—were regularly exposed to new paint materials and application techniques.[44] The Abstract Expressionists redefined painting with their monochrome expanses or wall-sized explosions of dripped, spattered, and poured paint. Although their goals were similar, the artists' individual styles were distinct: de Kooning's aggressive brushstrokes stand in stark contrast to Pollock's intricately dripped masses, for example; and the tiny, delicate marks at the edges of Rothko's luminous glazes (Crow 2005) seem unrelated to Still's tactile, opaque crusts of paint. "If the painting of de Kooning and Pollock may be called one of commission, in which the artist doesn't hesitate to reveal himself," observed Sam Hunter, "that of Still and Rothko is an art of omission, where more is suggested than stated" (1959, 154).[45] The groundbreaking work of Hofmann and his colleagues was at once alien and immediate, both abstract and expressive.

New paint materials do not define Abstract Expressionist painting, but the appearance of new materials in the practitioners' palettes led to changes in artistic practice that shaped modern art history. Modern paints appeared on the scene concurrently with the modern artists' search for a new means of artistic expression, and experimental paints encouraged experimental practices that affected the artists' relationship with both modern and traditional materials. "My opinion is that new needs need new techniques," proclaimed Pollock, who was one of the most well-known advocates of new paint materials, "and the modern artists have found new ways and new means of making their statements" (O'Connor 1967, 79).

In Abstract Expressionism, direct engagement with the materials of painting took precedence in art making. The work of the Abstract Expressionists was promoted in contemporary press as the "[direct] experience of paint and canvas" (Goodnough 1951, 60), and such rising artists as Guston, who considered himself "a very spiritual man, not interested in paint," now found himself suddenly "very physical and involved with matter" (1990). The modern canvas was no longer Greenberg's frontal, flattened plane, nor was it Rosenberg's arena stage; now it was also, as art historian Timothy J. Clark put it, "the intersection of body...and medium" (1999, 331).

Abstract Expressionist arts practice fundamentally changed the course of art criticism. The territorial disputes of Greenberg and Rosenberg have overshadowed an important commonality: both men's attention to the material aspects of the avant-garde. Despite latter-day claims by such critics as Rosalind Krauss that Greenberg intentionally sublimated his response to an artist's materials (1994, 243–48), from the start of his career, Greenberg drew attention to the relationship between materials and creativity. As early as "Avant-Garde and Kitsch," published in 1939, Greenberg proclaimed that modern painters "derive their chief inspiration from the medium [in which] they work" (1939, 36). Likewise, an artist's direct relationship with his materials played a primary role in Rosenberg's definition of the new trend in modern art. "The painter no longer approached his easel with an image in his mind," asserted Rosenberg (1952, 23); "he went up to it with material in his hand to do something to that other piece of material in front of him."

The material components of modern art also drew the interest of such contemporaneous arts journalists as *ARTnews* editor Thomas Hess, who spearheaded the magazine's series of more than eighty articles focused on modern artists' studio practices.[46] Contemporary explorations of "materiality" by such scholars as Yve-Alain Bois and Krauss (2000) and Richard Shiff (1987, 2011) have their origins in the new relationship between material and artist explored by the Abstract Expressionists. "In [t]his unapologetic materialism," wrote Hunter in the catalogue for a 1956 exhibition of works by Pollock at the Museum of Modern Art, "there [is] an effort to breathe spirit into the refractory matter that [is] the substance of [the] art" (1956–57, 12). Critics from Michael Fried (1964) to Donald Kuspit have positioned art criticism as a pivotal player in the development of modern art. The critic "makes the first, and presumably freshest, response to the work of art, grasps it when it is still new and strange, and gives us a preliminary hold on its meaning," said Kuspit. "Indeed, part of the critic's task is to win [modern art] a hearing at the court of history" (Foster 1980, ix). If this is true, it is Abstract Expressionism that is responsible for placing materials within the grasp of both the artists and the critics who fashioned modern art history.

Despite the new dynamic between artist and artwork, mid-twentieth-century artists nevertheless retained control over their materials.[47] While some of the radical painting techniques introduced by the Abstract Expressionists were facilitated by the appearance, surface qualities, and handling characteristics of new paint media, not all the artists were using new materials, and the artists' message was not determined by the materials themselves. "The medium becomes the work of art," said Hofmann, "only when the artist is intuitive and at the same time masters its essential nature and the principles which govern it" (Hofmann 1948, 59). Abstract Expressionism is the moment when process became a fundamental language in modern painting. In this art, claimed Rosenberg (1969, 116), "the artist became the medium of [his or her] medium."

The Abstract Expressionists initiated a paradigm shift in art practice from product to process that influenced subsequent generations of artists and changed the course of modern art. In his essay on the overt role of media in art produced on the heels of Abstract Expressionism, historian Thierry de Duve noted that "the medium in its specificity is not simply a matter of physical constituents; it comprises technical know-how, cultural habits, working procedures and disciplines—all the conventions of a given art whose definition is throughout historical" (1996, 210). Materiality (and conversely, immateriality) rose to prominence in the 1960s and 1970s, but as art historian David Anfam noted in his monograph on the movement, "scarcely any artist of [the next] generation could not have known about the signifying force of paint and touch that was central to Abstract Expressionism" (1990, 191). "The moment it leaves [Hofmann's] hand," claimed *New York Times* critic Brian O'Doherty (1963, 149), "the paint is alive and kicking."

The Abstract Expressionists' experimental approach to materials was an important influence on the art movements that followed. The lyrical abstraction that so captivated Greenberg's later criticism, for example, was a direct progression from Abstract Expressionism. "Taking hints from [Pollock's] methods and materials," recalled stain painting innovator Frankenthaler, "I experimented and proceeded to try other ideas" (Glueck 1977, 85; V. Mecklenburg 2008, 131). Emerging artists with an interest in materials drew inspiration from the Abstract Expressionists. "Both Pollock and Hofmann seemed to me to have solved the problem," noted Stella. "I didn't have to go all the way back and worry again about where I stood in relation to Matisse and Picasso. I could worry about where I stood in relation to Hofmann and Pollock" (de Antonio and Tuchman 1984, 84). Reinhardt's search for "color that gives off light"[48] foretold Flavin's rooms literally filled with light, just as Newman's monochrome expanses made possible sculptor Richard Serra's bending of sculptures around the viewer—and enabled that viewer's ability to step directly into the earthworks of Walter De Maria and Robert Smithson. Similarly, as Rosenberg and Sandler saw the seeds of performance art in Pollock's attempts to "literally be in the painting" (Sandler 1970, 102), the trajectory from such artists as Pollock and Hofmann leads to the body-as-instrument work of Yves Klein, Janine Antoni, and Kazuo Shiraga,[49] as well as to the body-art performances of such artists as Vito Acconci, Carolee Schneemann, and Karen Finley. In 1924 abstract painter Paul Klee told an audience: "Nobody would affirm that the tree grows its crown in the image of its root. Between above and below can be no mirrored reflection. It is obvious that different functions expanding in different elements must produce vital divergencies.... And yet, standing at his appointed place, the trunk of the tree...gather[s] and pass[es] on what comes to him from the depths" (Klee 1966).

The technical study of an artist's work has potential far beyond an increased understanding of paint mechanics and degradation; technical studies of mid-twentieth-century art also illuminate a pivotal shift in the relationship between materials and art practice. The Abstract Expressionists' relationship with their materials changed modern art. Scholars who place mid-twentieth-century painting along a simple figurative–nonfigurative trajectory miss the important link between contemporary arts practice and its roots in the shifting dialogue between artist and art making found in the work of Abstract Expressionist painters. It would be shortsighted to insist that the history of modern material culture begins when the materials have taken center stage. A prominent role in the perpetual evolution of modern art should be assigned to the pioneering Abstract Expressionists who, as historian Michael Auping noted, "disregarded the idea of traditional categories and focused on the matter itself and the various procedures suggested by different materials" (1987, 152).

Materials in Hofmann's Classrooms

Hofmann's classrooms in New York and Provincetown were places of active dialogue between art manufacturers and practitioners as experimental media entered mid-twentieth-century art practice—fertile environments for both students and colleagues to examine and influence the world around them. "To avoid being academic," Hofmann claimed in a 1937 brochure, "a school of art must be a vital participant in contemporary aesthetics" (Hofmann papers, [ca. 1904]–78). As was his school in Munich, Hofmann's American schools served as a bridge between the experimental spirit of his progenitors and that of his descendants. The painting styles and techniques of émigré modernists and American innovators alike all passed through Hofmann's schools, which were recalled fondly by students as places "where such ideas were aired, tried and discussed" (Dickey 1997, 82).

Hofmann's classrooms were a crossroads for information about advancing paint technology. The supply list for students attending his early Provincetown classes, for example, combined the traditional and modern paint formulations popular during his years in Europe with paints recently developed by innovative American paint manufacturers.[50] The titanium-zinc Permalba white of Philadelphia paint maker F. Weber and Company (now Martin F. Weber Co.) was listed as a required pigment, for example, and oil paints made by the nascent American manufacturer Permanent Pigments (established in 1933) were recommended alongside those of established European manufacturers including Winsor and Newton, Rembrandt, and Blockx (Hofmann papers, [ca. 1904]–78; Kuper collection, 1932–64). Hofmann increased his students' exposure to developing paint technology through the classroom visits of paint makers Leonard Bocour of Bocour Artist Colors and Henry Levison of Permanent Pigments (Marontate 1996, 225; Dickey 2012) and through information provided by fellow Hofmann students in leadership positions at institutions experimenting with new paint media. Current and former Hofmann students in positions of authority within the WPA during this period included Glenn Wessels and Cameron Booth in the San Francisco region (Wessels as Bay Area technical supervisor and Booth as a supervisor in Oakland) and Harry Holtzman and Burgoyne Diller in New York (where they shared a supervisory post in the Federal Art Project's mural division) (Lane 1984, 28; Dickey 2011, 174). Hofmann's art schools were built on modernist ideals of exchange and community that encouraged the transfer of information about avant-garde materials and procedures. "He didn't so much create painters," recalled Krasner, "as a milieu in which painting could exist" (*New York Times* 1966, 33) (fig. 2.8).

Hofmann's lectures on the material nature of painting were widely influential. They played a pivotal role in the work of such critics as Rosenberg and Greenberg, whose focus on art materials and art practice changed the trajectory of modern art criticism, as discussed above. "Hofmann has not yet published his views [on modern art]," Greenberg would comment (1945, 469), "but they have already directly and indirectly influenced many, including this writer—who owes more to the initial illumination received from Hofmann's lectures than to any other source." The relationship between artist and material was also fundamental to the theories of Rosenberg, Greenberg's ideological opponent, who placed Hofmann among the first of his "action" painters.[51]

Hofmann's popular late-career works—most produced well after his seventieth birthday—are pulsing networks of color that at once recall and transcend his influences;

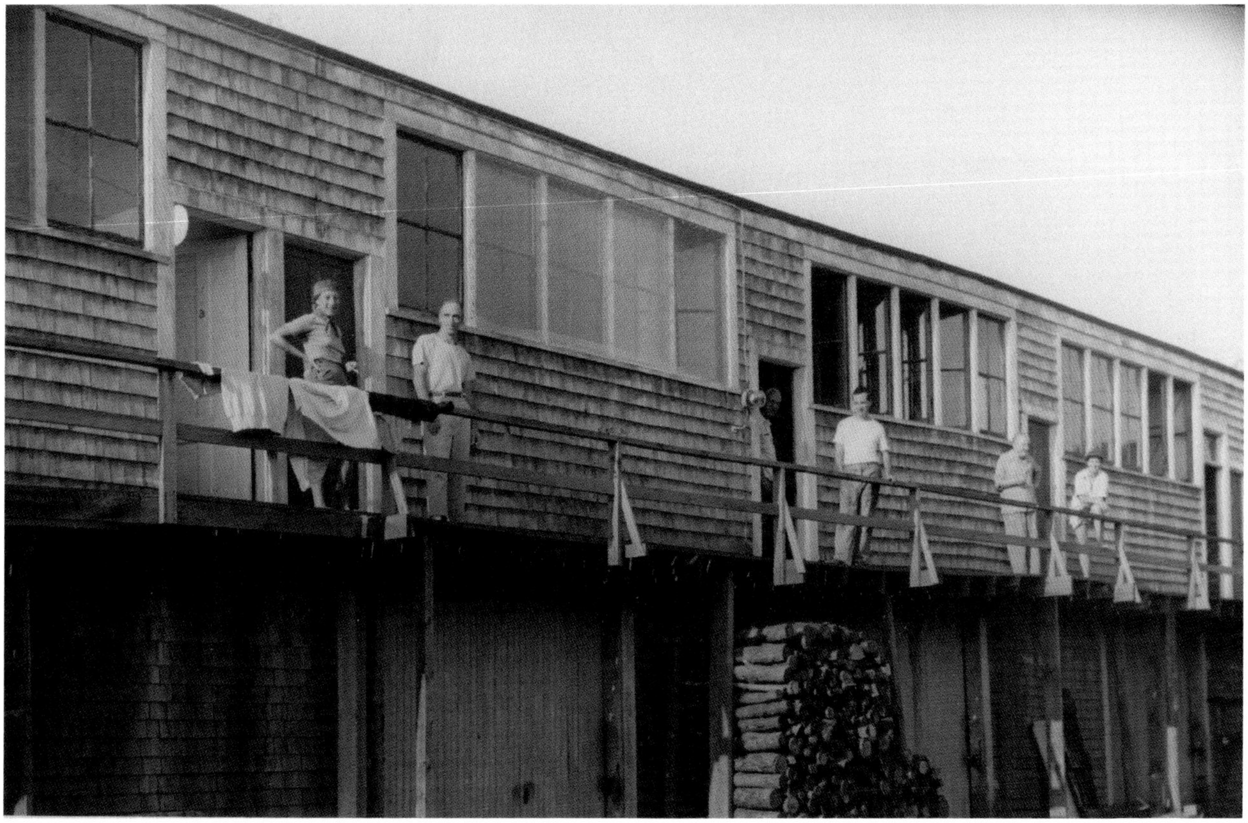

FIGURE 2.8
The catwalk at Days Lumberyard, Pearl Street, Provincetown, ca. 1953. Left to right: Elaine de Kooning, Charles Egan, Sam Kootz, Fritz Bultman, Hans Hofmann, and Perle Fine.
Photo: Maurice Berezov.

they are physical plays on paint and light that are simultaneously referential and singular. "Whether Hofmann splashes paint onto the canvas, or brushes it in heavy impasto," proclaimed Peter Selz, former Museum of Modern Art curator and founding director of the University Art Museum of the University of California, Berkeley, "he turns the surface into a seemingly living witness to his manipulation of paint" (1993, 130). Indeed, the technical study of Abstract Expressionist materials can do no better than to look to Hofmann for access to the zeitgeist of this period in modern American painting.

3 Hofmann's Late-Career Paintings

> Creation is dominated by three absolutely different factors: first, nature, which affects us by its laws; second, the artist who creates a spiritual contact with nature and with his materials; and third, the medium of expression through which the artist translates his inner world.
> — HANS HOFMANN, "Painting and Culture," 1931

> A thing in itself never expresses anything. It is the relation between things that gives meaning to them.
> — HANS HOFMANN, 1948

Hofmann's lifelong interest in the nature of existence guided his journey from literal to experiential representation in art and is reflected in an intentional consistency in the artist's late works. The paintings produced by Hofmann from the mid-1950s onward are the most acclaimed of the artist's work. Critical and scholarly attention is commonly focused on the "signature" works of this late period that feature boldly colored rectangles, but discussion of these compositions has thus far been limited to literal interpretations of the geometric forms. The symbolic nature of these forms becomes apparent when they are viewed in the context of the breadth of Hofmann's late-career oeuvre. The appearance and evolution of the artist's colored rectangles highlight an underlying theme of depth and complexity that ties together seemingly disparate compositions. Hofmann's rectangles allude to—but do not define—the artist's signature endeavor.

Origins and Development of Hofmann's Late-Career Style

The colored rectangles most people associate with Hofmann did not emerge in the artist's work until the mid-1950s, shortly before he retired from teaching. Hofmann's late-career style disassembled and recombined the influences of his modernist training into a completely new visual vocabulary. The artist advanced the studies of his Cubist and Fauvist forebears by using the relational effects of neighboring colors to create dimensionality, a phenomenon Hofmann called "push and pull." "Push and Pull is a colloquial expression," Hofmann explained in his teaching manual *The Painter and His Problems*, "applied for movement experienced in nature or created on the picture surface to detect the counterplay of movement in and out of depth" (1963b). Hofmann's signature style is distinguished by the primacy of relational color; in melding the structural planes of

Cubism with the symbolic color of Fauvism, Hofmann's compositions are driven by "symbolic structure"—color that shapes and energizes the picture plane. "Much as [Hofmann] relied on Cubist design throughout his career," critic Irving Sandler noted (1970, 138–39), "he was one of the first to loosen it by opening its closed planes and by using color in itself to determine structure.... He invested each area with a suggestion of the third dimension that made it feel ample." Hofmann constructed his compositions from planes of color representing "the space *in front* of the object, the space *in* the object, and the space *in back* of the object" (1932–33, 4), but maintained that the relationship of these planes to one another and to other elements of the composition was ultimately determined by the relationship between colors. "It is the color *development*," Hofmann emphasized, "that determines the form" (1955, 3).

The breakthrough in Hofmann's use of color planes emerged concurrently with the artist's work in mosaic. In 1955 art dealer Samuel Kootz arranged for Hofmann to create a mosaic mural for the lobby of 711 Third Avenue, a building by architect William Lescaze and builders William Kaufman and Jack D. Weiler (fig. 3.1). This project was the first of the artist's mosaic designs to be produced,[52] and the experience of translating a sketch to a full-size mosaic directly influenced Hofmann's work on canvas.

Hofmann enlisted the help of students, including James Gahagan and Max Spoerri, to translate the Third Avenue sketches into a full-size design plan. During the 1956 phase of mural preparations, painted papers used by Gahagan as place markers (fig. 3.2) caught Hofmann's attention and became permanent components of the mural design. "One morning [Gahagan] painted up a pile of rectangle sheets of paper and laid them flat on the mural [plan]," Tina Dickey reported. "Hofmann arrived and began to move them about with excitement. Once they established the rhythmical structure of that section, Gahagan expected to pick up the papers to fill in the area with paint. But Hofmann

FIGURE 3.1
Hans Hofmann, mosaic mural at 711 Third Avenue, New York, 1956.

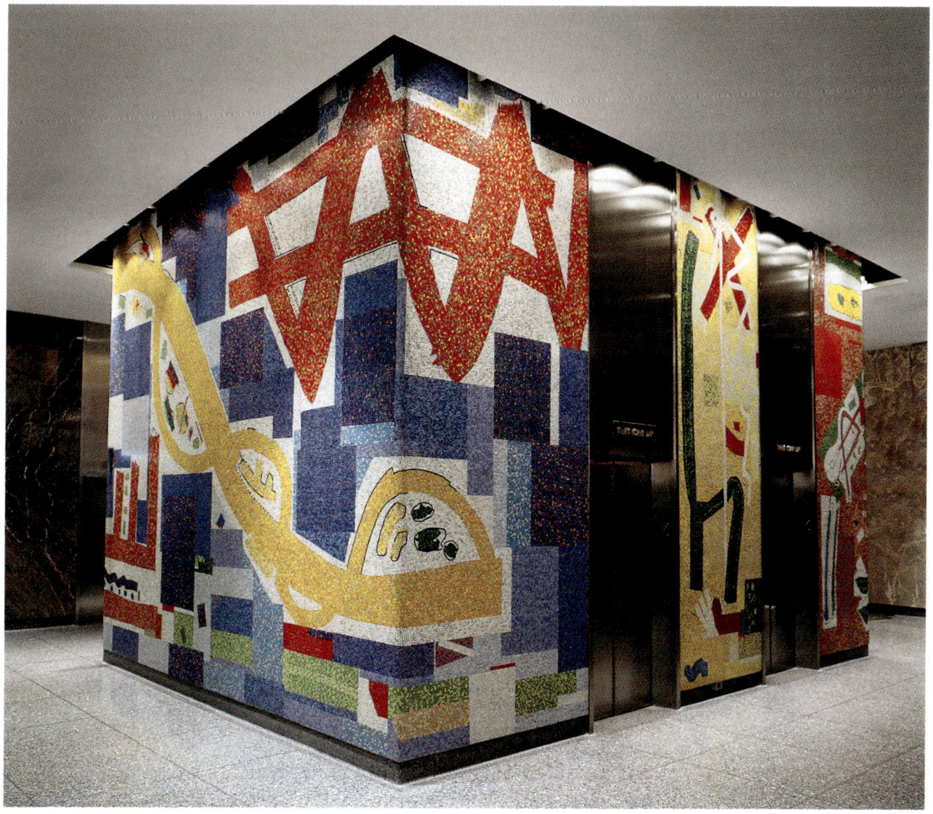

FIGURE 3.2
William Freed, Max Spoerri, and James Gahagan preparing the design for the Third Avenue mural, 1956.

Photo: Gene Lesser.

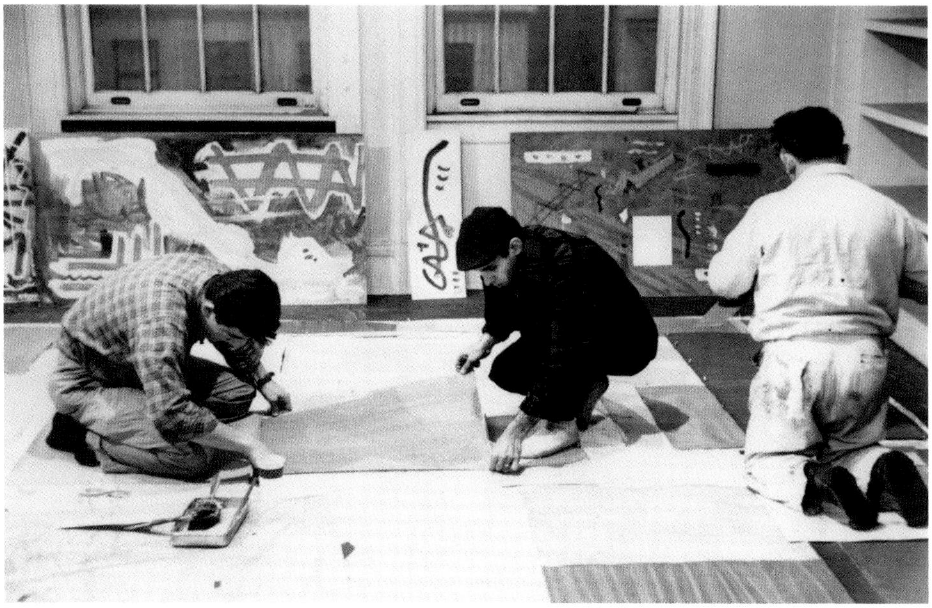

insisted they remain part of the composition" (Dickey 2002, 274).[53] Hofmann had previously torn and rearranged portions of student drawings to demonstrate the effect of shifting compositional planes, but it was around the time of the Third Avenue project that blocks of solid color became a central presence in Hofmann's teaching. According to Dickey, "During the summer 1956 session, at one Friday critique in Provincetown, seventy people crowded close in rapt attention.... Gahagan stood by a table full of colored paper rectangles. Haynes Ownsby had never seen big rectangles of colored papers used in a Friday critique before" (Dickey 2011, 317).

In 1956 free-floating planes of color appeared in Hofmann's paintings. Critical inquiry has acknowledged similarly shaped elements in Hofmann's earlier compositions—"the rectangles started poking through the Cubist space of Hofmann's paintings around 1954...but gained no solid purchase," according to artist-historian Walter Darby Bannard (1969, 39)—as well as the sudden transformation of colored forms into fundamental components of the artist's late work: "solidified rectangles first figure as a major compositional element in 1956, and from this time on, [Hofmann] was irrevocably committed to their use," noted historian Cynthia Goodman (1986, 70). But this commentary appears without reference to the transformative link provided by the colored papers used in Hofmann's mural projects. The link between Hofmann's murals and his signature compositional elements remains largely ignored, despite direct physical evidence of the mural projects' impact on Hofmann's late-career paintings.

The colored papers Hofmann used in laying out his mural designs were enlisted in composing the artist's late works on canvas, and anecdotal and photographic records of Hofmann at work document colored papers pinned to works in process (fig. 3.3). Evidence exists that Hofmann continued to utilize commercial and homemade color swatches as compositional aids for the rest of his career (Spoerri 2011). An inventory of the artist's studio contents following his death includes a list of commercially available "Color-aid" papers, a "portfolio of 100 excellent pieces of construction paper," and "several hundred color blends [by] Mr. Hofmann on 3 × 5 cards" (Hofmann papers, [ca. 1904]–78).

Hofmann's use of rectangles should not be confused with the hard-edge figures employed by contemporaneous nonobjective painters. Hofmann's signature compositions employ abstract figuration tied to objects in the natural world, while such works

FIGURE 3.3
Hofmann in his Provincetown studio, 1962.

Photo: Arnold Newman

as the *Homage to the Square* series produced by former Hofmann student Josef Albers are devoid of corporeal subject matter. Albers replicated a set series of compositions and measured and recorded his color formulations like laboratory experiments, in stark contrast with the emotion and empathy at the core of Hofmann's relationship to color and form. "[His] color has to do with feelings" scholar William Agee noted of Hofmann's late work, "and Hofmann is nothing but feelings" (2012, 7).

Similarly, while Hofmann and his wife, Maria Wolfegg, stored dozens of early, nature-based abstract works by nonobjective pioneer Wassily Kandinsky in his Munich studio,[54] any potential expression of his fellow Münchner's influence is quickly subsumed into Hofmann's resolutely representative compositions. The moment Hofmann's color planes move from the wall to the canvas, they become aspects of the artist's established personal vocabulary—abstractions of form with direct ties to identifiable subject matter. If Hofmann's rectangles bear evidence of nonobjective experimentation, that influence is like all of Hofmann's formative experiences—interpreted and advanced at the hand of the master modernist.

Formal analysis of Hofmann's color planes reveals the underlying continuity in the artist's late-career work. Terms previously used to describe the planes—from Goodman's "substantial slabs of color" (1986, 86) to Bannard's "large-scale small-piece all-over Cubism [LASPAC]" (1969, 38)—assign the planes specifics of appearance or function

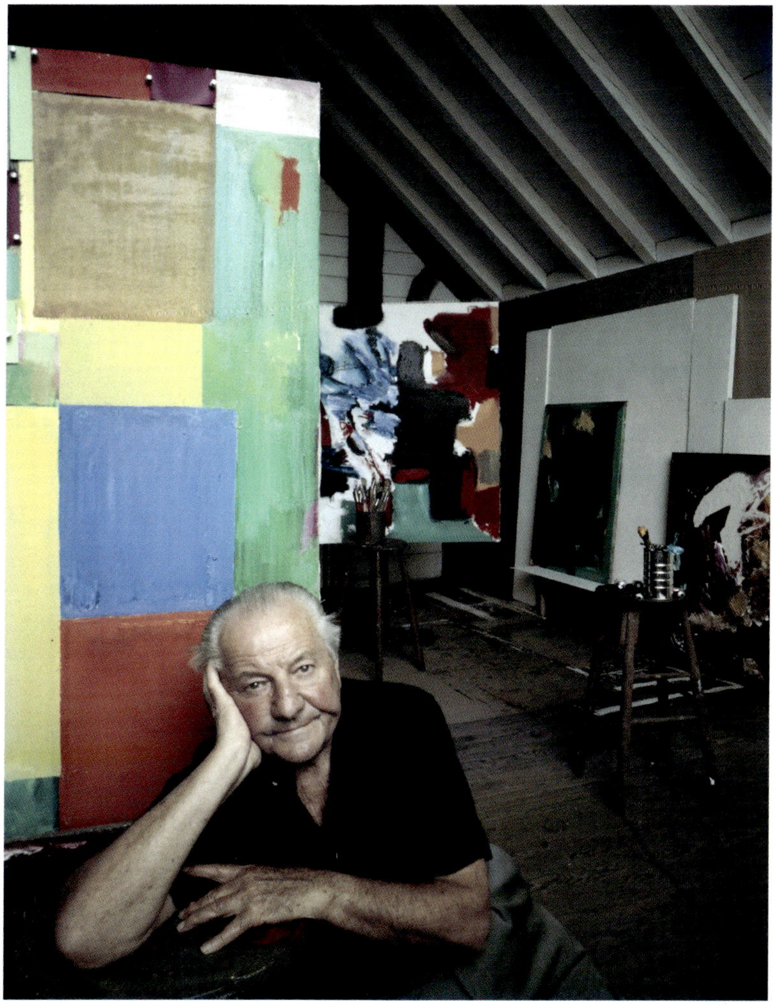

FIGURE 3.4
Hans Hofmann, *Combinable Wall I and II*, 1961. Oil on canvas, 214.6 × 285.8 cm (84½ × 112½ in.). BAM/PFA, 1963.10. Gift of Hans Hofmann.

FIGURE 3.5
Hans Hofmann, *Memoria in Aeternum*, 1962. Oil on canvas, 213.4 × 183.2 cm (84 × 72.1 in.). New York, Museum of Modern Art, 399.1963. Gift of the artist.

These paintings are examples of overtly figured color planes presented in closed (left) and open (right) compositions.

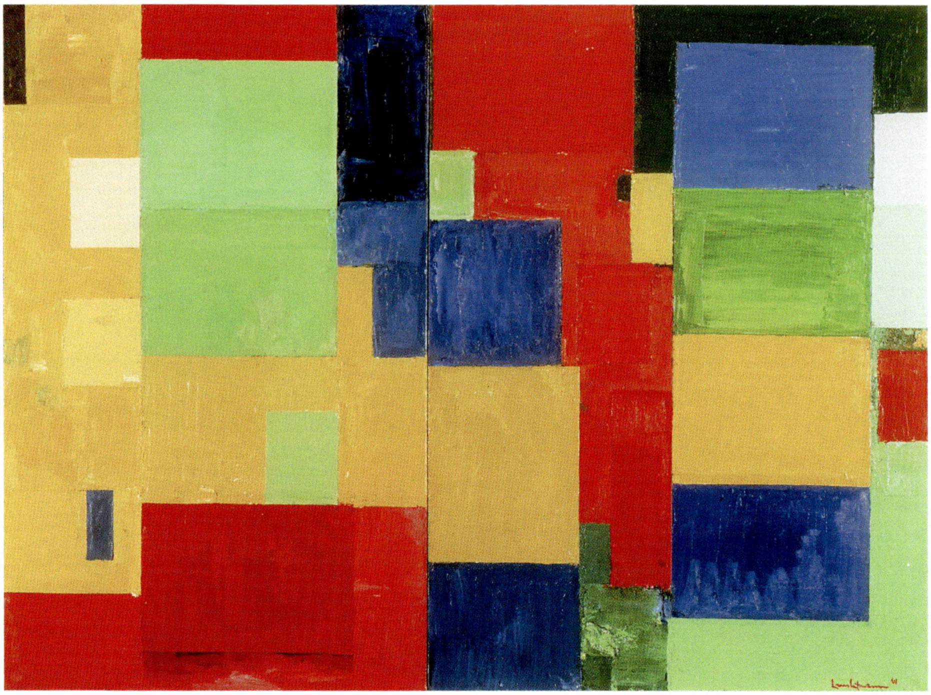

that limit their inherent flexibility. As noted above, Hofmann's compositions are driven by relational color and are perhaps more appropriately called "color planes." Formal analysis of the late paintings sheds light on the evolving function of the color planes and, in doing so, reveals the theme running throughout the artist's late oeuvre.

Each of Hofmann's late paintings can be sited along two abstract, formalist trajectories: open/closed composition, and overt/obscure figuration (the figures in this case being Hofmann's abstracted color planes). The first trajectory is defined by the concentration of color planes, and the second trajectory is defined by the physical boundaries of those planes. The 1961 diptych *Combinable Wall I and II* (fig. 3.4), for example, is a closed composition, filled with interlocked planes of heavy paint through which no open space is visible. *Combinable Wall I and II* also represents overt figuration in the presentation of the color planes. The boundaries of each color plane are clearly defined, the thick paint of each plane having been applied with a palette knife to achieve a hard-edge border that gives physical presence to these primary compositional elements. "[These] planes are paint loaded," noted Sandler. "The vivid, dissonant colors . . . strain to burst the stable rectangular containers" (1970, 144). In contrast, Hofmann's 1962 painting *Memoria in Aeternum*—dedicated to the memory of Arthur Carles, Arshile Gorky, Jackson Pollock, Bradley Walker Tomlin, and Franz Kline—combines open composition with overt figuration, floating a handful of hard-edge planes over an open area of brushy color (fig. 3.5).

Hofmann's open compositions typically feature large areas of exposed ground and brushy color that support a handful of thickly painted color planes or splashes of heavy paint. In *Memoria in Aeternum,* two glowing color planes of the style seen in *Combinable Wall I and II* float within a murky background that partially obscures remnants of luminescent color atop a bright white ground. "I cannot think of another painting that intimates immortal hopes by such strictly abstract means," proclaimed modern art critic Harold Rosenberg, who said the painting gave "the impression of being inside the earth" (1963b, 104).

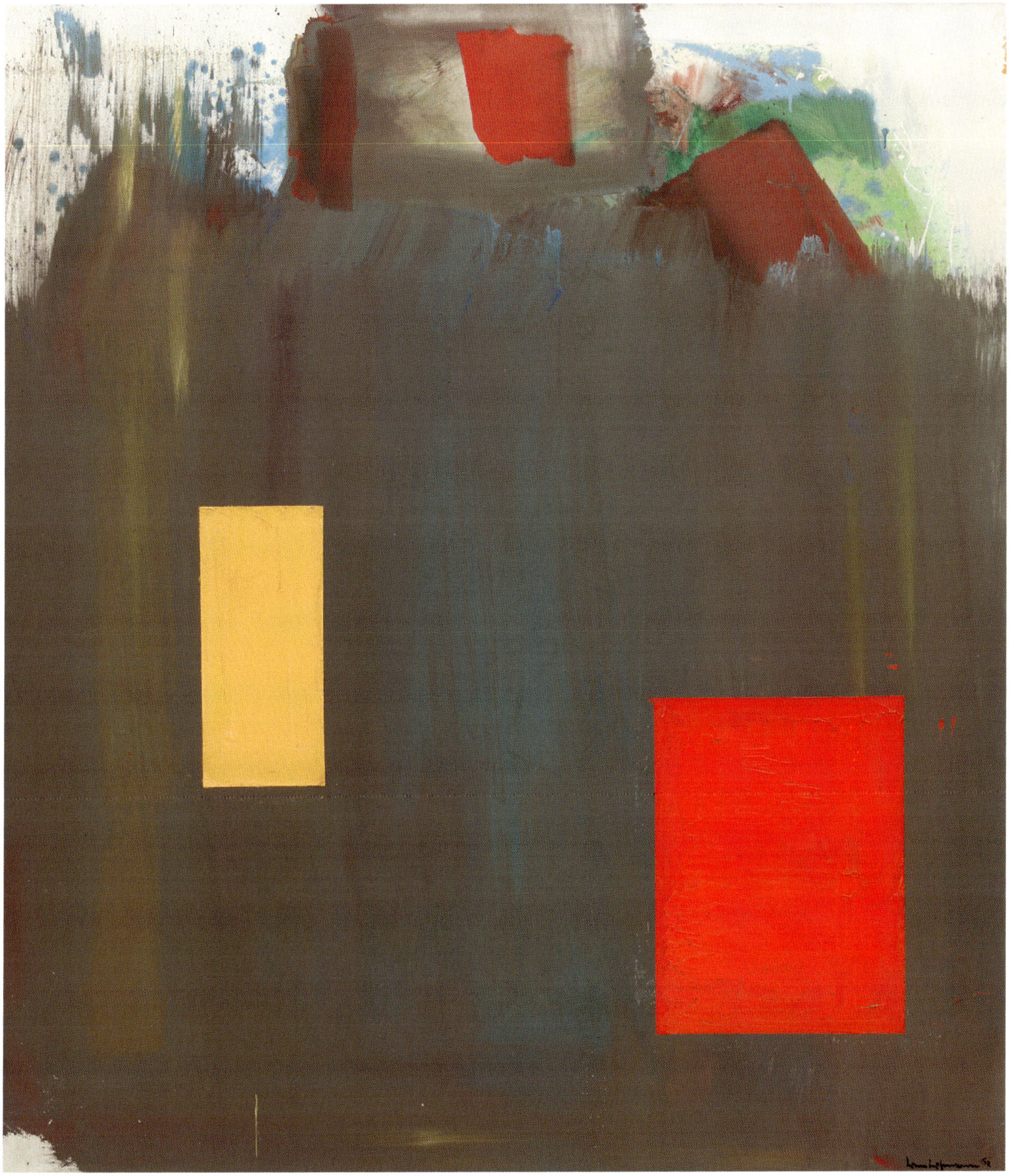

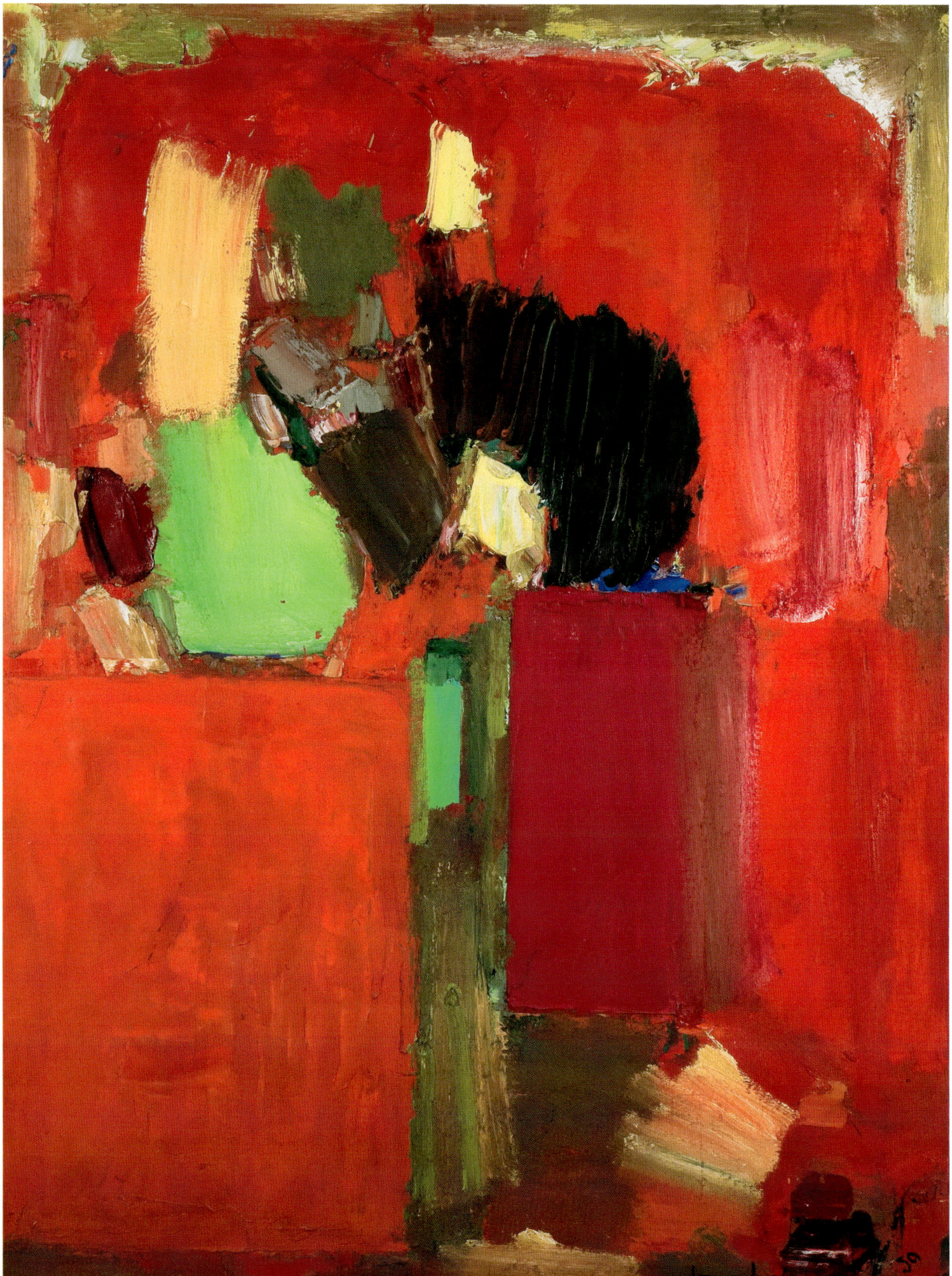

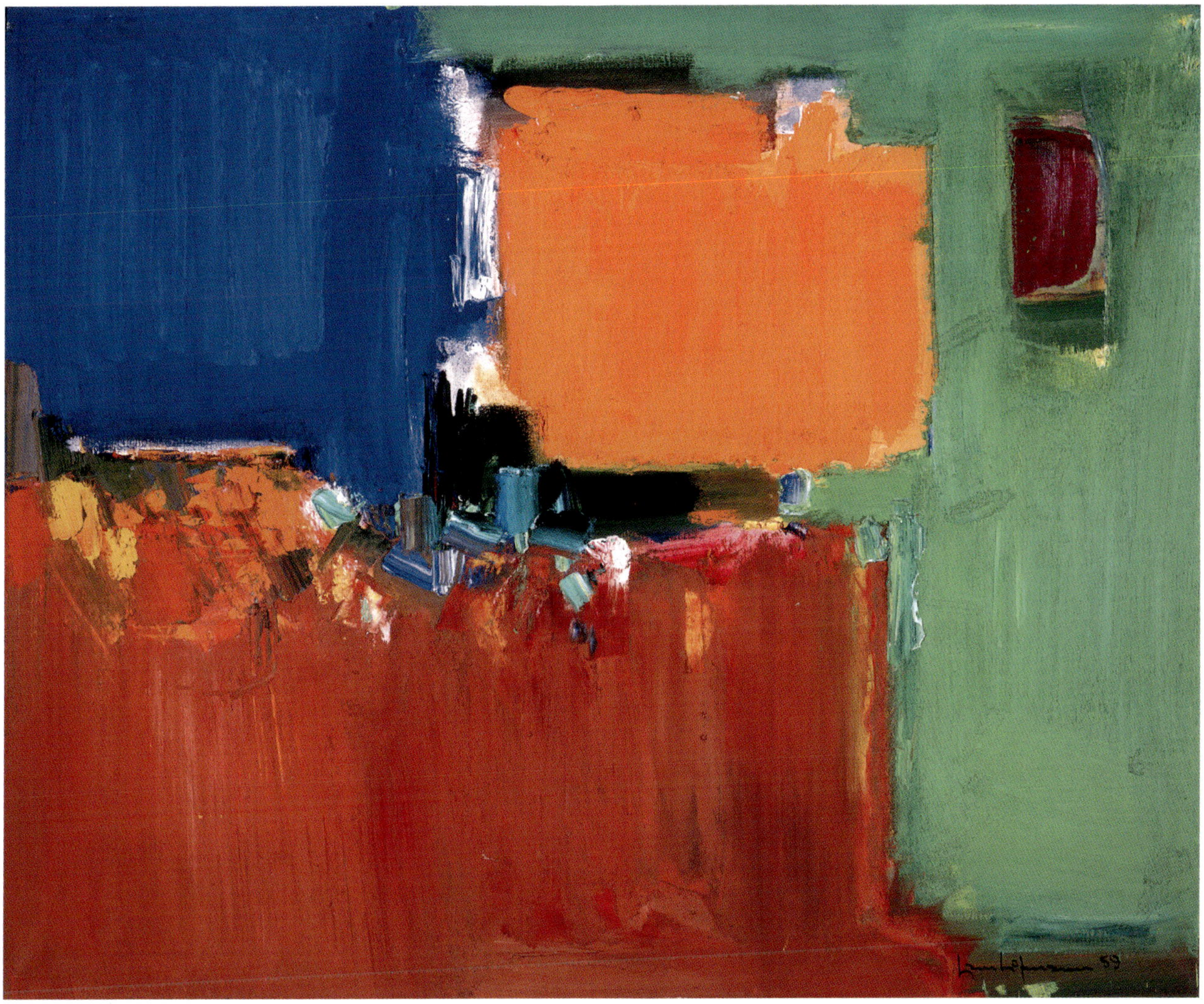

FIGURE 3.6
Hans Hofmann, *Ruby Gold*, 1959. Oil on canvas, 140.7 × 102.9 cm (55⅜ × 40½ in.). Rochester, New York, Memorial Art Gallery (MAG), 60.37. Marion Stratton Gould Fund.

FIGURE 3.7
Hans Hofmann, *Indian Summer*, 1959. Oil on canvas, 152.7 × 183.4 cm (60⅛ × 72¼ in.). BAM/PFA, 1965.11. Gift of Hans Hofmann.

These paintings are examples of obscurely figured color planes in closed compositions.

Color planes with organic shapes or indistinct borders also appear in both closed and open compositions. Hofmann's obscurely figured compositions range from the brushy-edged rectangles packed into the closed compositions of *Ruby Gold* and *Indian Summer* (both 1959; figs. 3.6 and 3.7) to the floating bursts of paint that punctuate the open expanses of *Tormented Bull* (1961; fig. 3.8) and *Struwel Peter* (1965; fig. 3.9). Obscurely figured planes do not always fall parallel to the picture edge,[55] but they remain leading characters in the color relationships that anchor and activate Hofmann's compositions.

Other works fall between the formalist polarities described above. Obscurely figured splashes of paint, for example, appear with overtly delineated color planes in the 1959 painting *The Vanquished* (fig. 3.10), and soft-edged rectangles mix with brushy paint strokes to envelop more-geometric shapes in *The Clash* (1964; fig. 3.11).

Although Hofmann continued to produce painting along these formalist spectra throughout his career, there is a distinct progression in the artist's late work from closed to open compositions, as the closed works of the late 1950s (including *Ruby Gold, Indian Summer,* and *The Vanquished*) give way to a preference for open compositions in the early 1960s (such as *Memoria in Aeternum, The Clash,* and *Struwel Peter*). 1961 is a transitional year, featuring both the closed/overt *Combinable Wall I and II* and the open/obscure *Tormented Bull.*

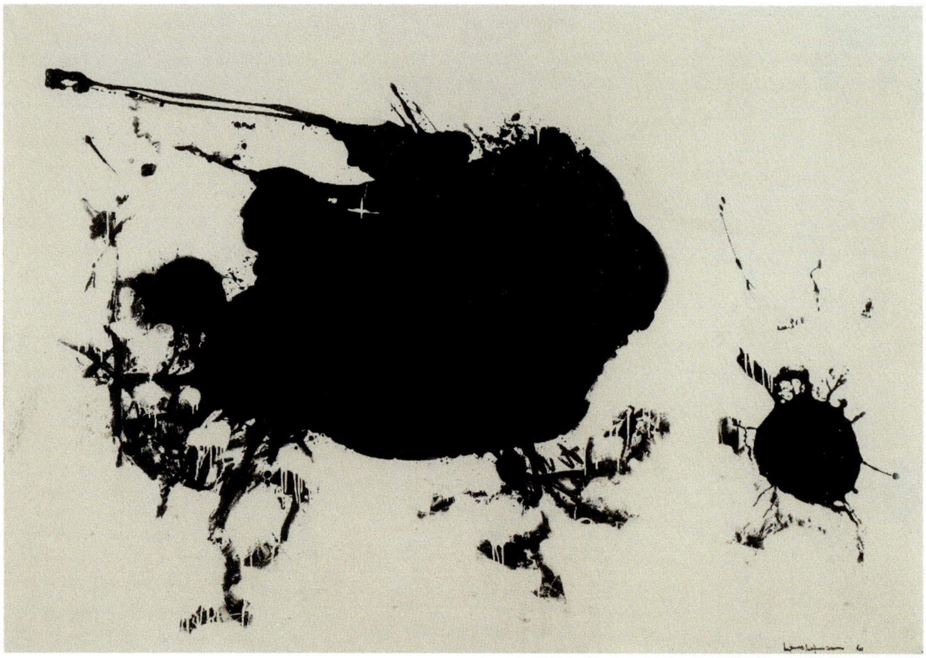

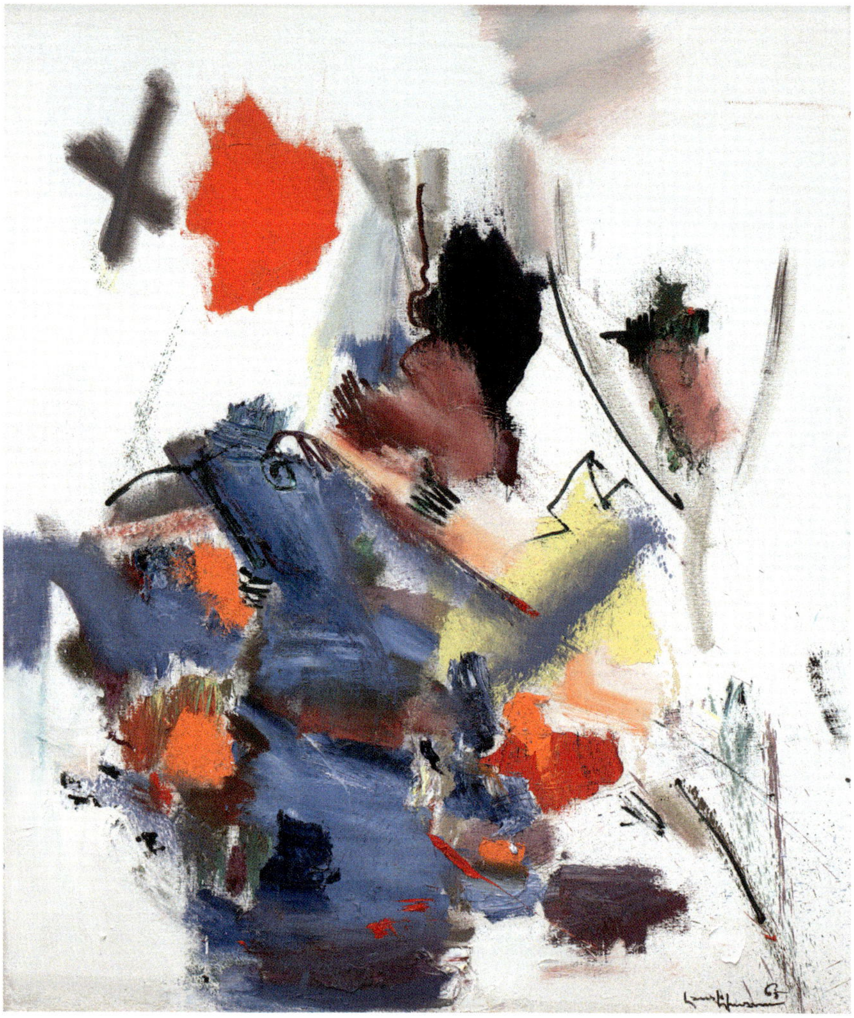

FIGURE 3.8
Hans Hofmann, *Tormented Bull*, 1961. Oil and enamel on canvas, 152.7 × 214 cm (60⅛ × 84¼ in.). BAM/PFA, 1963.6. Gift of Hans Hofmann.

FIGURE 3.9
Hans Hofmann, *Struwel Peter*, 1965. Oil on canvas, 183.2 × 153 cm (72⅛ × 60¼ in.). BAM/PFA, 1966.5. Gift of Hans Hofmann.

These paintings are examples of obscurely figured color planes in open compositions.

FIGURE 3.10
Hans Hofmann, *The Vanquished*, 1959. Oil and enamel on canvas, 91.8 × 122.2 cm (36⅛ × 48⅛ in.). BAM/PFA, 1966.49. Bequest of the artist.

FIGURE 3.11
Hans Hofmann, *The Clash*, 1964. Oil on canvas, 132.4 × 153 cm (52⅛ × 60¼ in.). BAM/PFA, 1965.8. Gift of Hans Hofmann.

These images represent a mixing of formalist approaches often seen in Hofmann's late-career work.

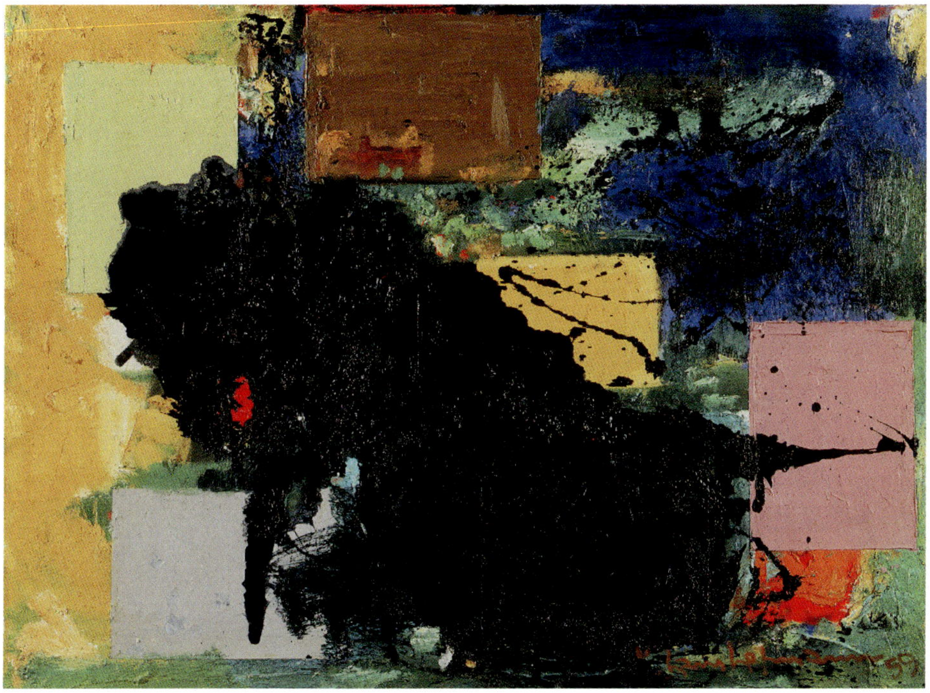

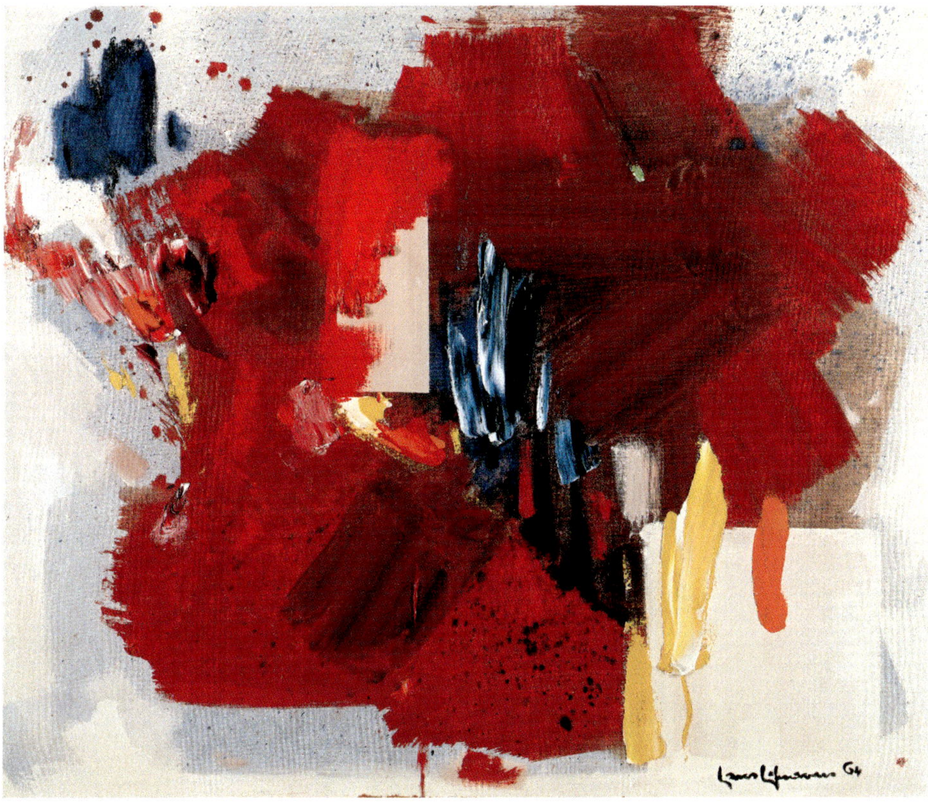

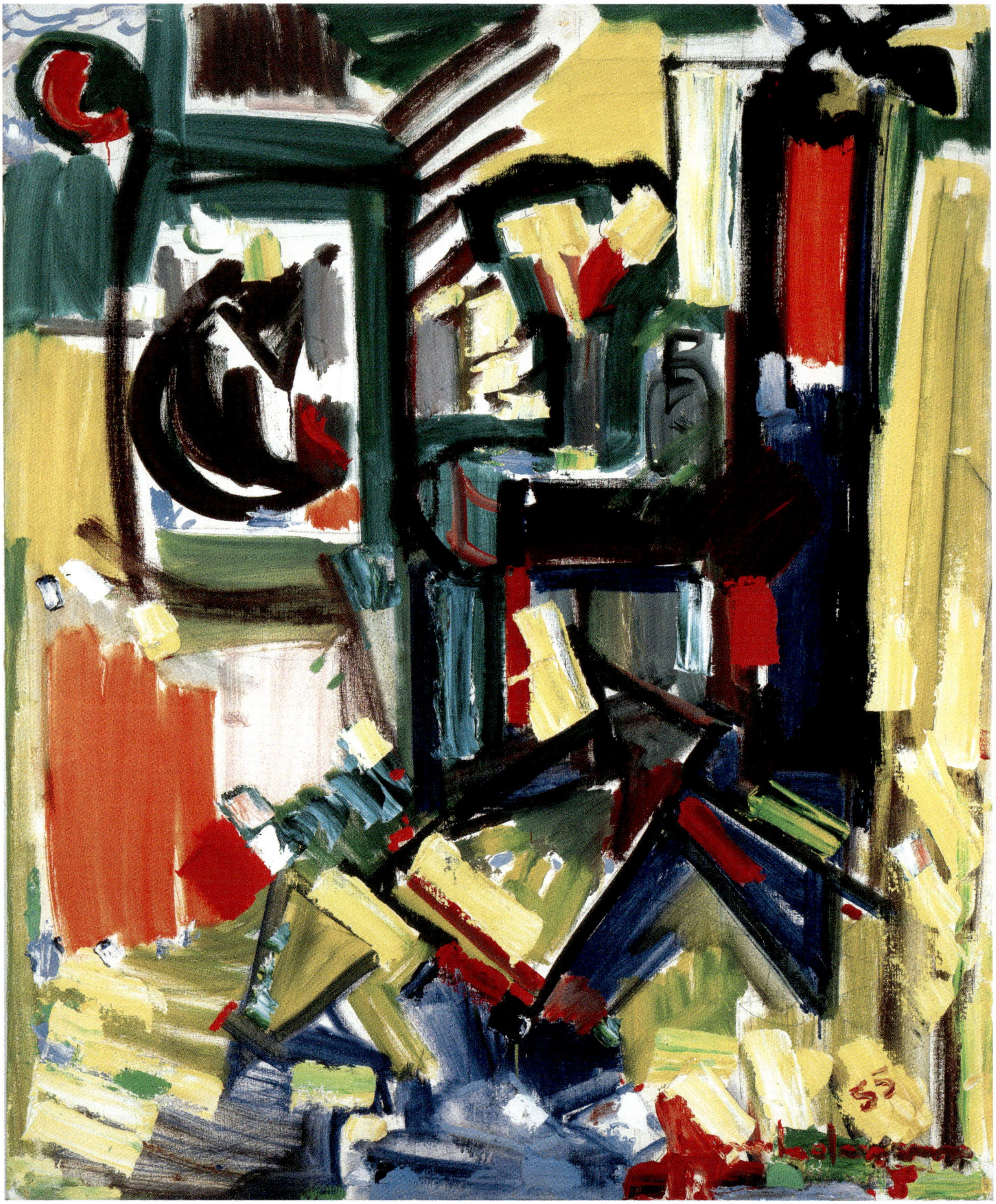

FIGURE 3.12
Hans Hofmann, *Exuberance*, 1955. Oil on canvas, 127 × 101.6 cm (50 × 40 in.). Buffalo, New York, Albright-Knox Art Gallery (AKAG), 1955:8. Gift of Seymour H. Knox Jr.

Despite the formal range of Hofmann's compositions, there is a consistency in the artist's drive to explore and distill his subject matter. The exploratory thread connecting Hofmann's work was missed by contemporaries such as art historian Peter Selz, who claimed that "unlike most of the New York painters... Hofmann did not work in series nor cultivate a single, signature style" (1993, 130). The variations of composition seen in Hofmann's late works are more reflections of changing emotion than of changing purpose. "As an artist I must conform to my nature," Hofmann told colleagues at the Studio 35 sessions. "My nature has a lyrical as well as a dramatic disposition" (Motherwell, Reinhardt, and Karpel 1951, 21). These formalist variations should not be confused with a lack of focus on the part of the artist. Hofmann's late work is a distillation of methods and materials in search of the essential reality behind his subject matter. "Artistic creation," according to Hofmann, "is the metamorphosis of the external physical aspects of a thing into a self-sustaining spiritual reality" (1948, 40). In this sense, even the most obscurely figured colors function as color planes in Hofmann's work, actively participating in the relational environment at the heart of his paintings. "There's not always a rectangle to be found," Bannard noted, "but... when physically absent they [leave] the imprint of their authority" (1969, 39).

In the last year of his life Hofmann took a significant step forward in his compositions. After nearly a decade spent exploring the energetic relationships between objects, he produced a handful of works that depict the forces at play within energy itself. In the first phase of this journey, his color planes reflect and then advance Cubist principles. The still-life origins of Hofmann's 1955 painting *Exuberance* (fig. 3.12) can be seen in the artist's static blocks of color, but just two years later, the artist moved beyond the literalness of Cubist composition to illustrate relationships between objects and their surroundings in works such as *Sommernachtstraum* of 1957 (fig. 3.13). The color planes in this work still perform a representative function but are no longer tied to the physical contours of the object. Hofmann asserted that his subject had not been abandoned but had instead been transformed. "I have never given up on the *object*," Hofmann said in 1957, the year he painted *Sommernachtstraum*. "When you analyze nature in regard to a picture, then the object is absorbed through light, or shadow, or color" (Wight 1957, 23). In *Sommernachtstraum* Hofmann's color planes emerge as symbolic descriptors of an object, as noted by curator William Seitz in his catalogue essay for the artist's 1963 retrospective at the Museum of Modern Art. "Hofmann is attentive to physical bodies of all categories, from mountains and buildings to still life and the posed model, but he sees them simplified and essentialized," Seitz said. "Following this interpretation, even Hofmann's most abstract paintings can be seen as a dialogue between stripped 'objects'" (1963, 11). Hofmann's metaphysical underpinnings are clear. "The artist's technical problem is how to transform the material with which he works back into the sphere of the spirit," Hofmann noted. "This two-way transformation proceeds from metaphysical perceptions... the search for the essential nature of reality. And so artistic creation is the metamorphosis of the external physical aspects of a thing into a self-sustaining... reality. Such is the magic act which takes place continuously in the development of a work of art" (1948, 40).

Sommernachtstraum is an early example of what can be called Hofmann's essentialist portraiture, compositions that explore the essence of their subject matter. In *Sommernachtstraum* Hofmann depicts the energetic relationships between object and environs through the relational energies of neighboring colors. "The creative process lies not in imitating, but in paralleling nature," wrote Hofmann, "[in] translating the impulse received from nature into the medium of expression through which the artist translates his inner world" (1931, 5).

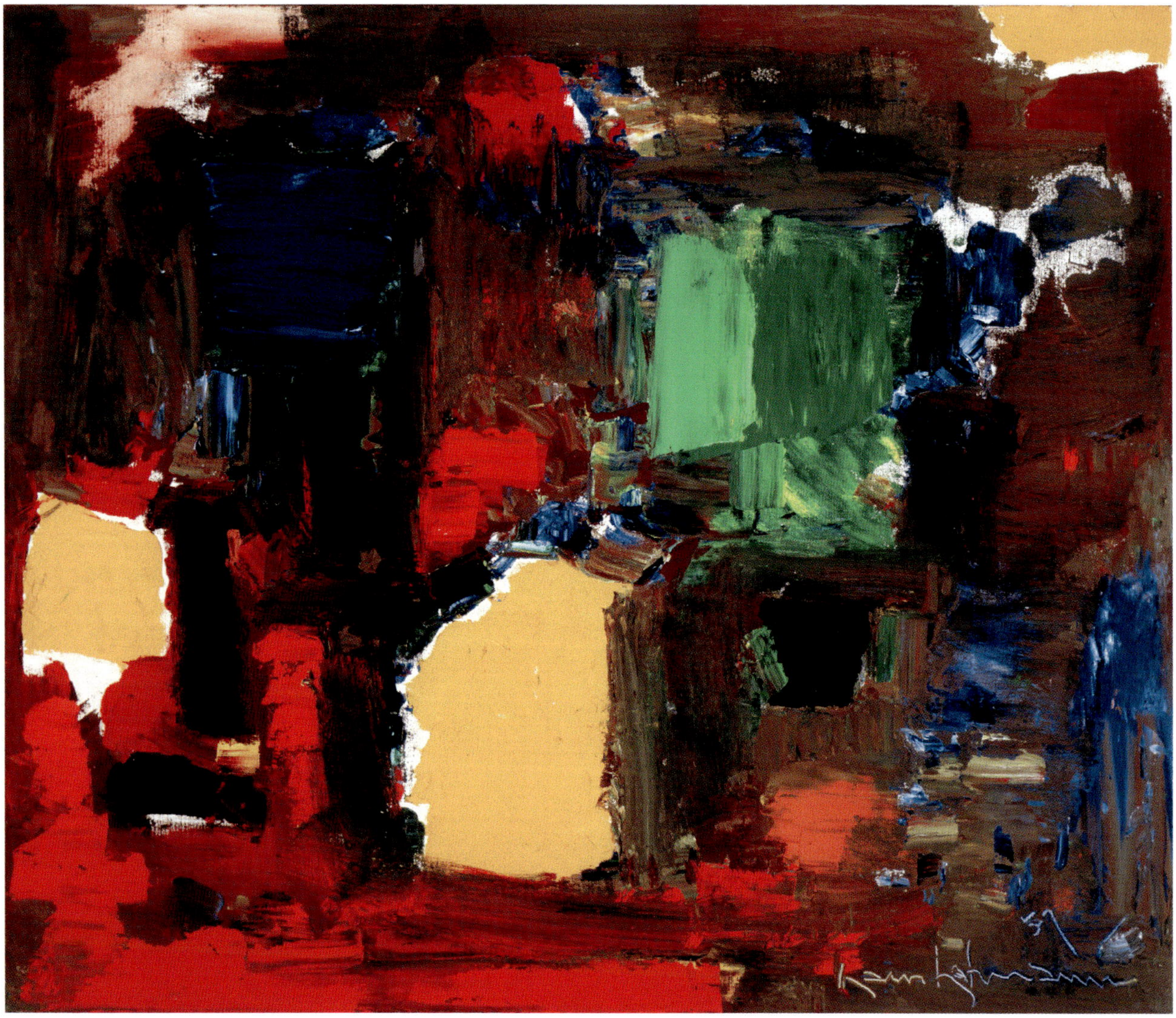

FIGURE 3.13
Hans Hofmann,
Sommernachtstraum, 1957.
Oil on canvas, 132.1 × 152.4 cm
(52 × 60 in.). AKAG, 1958:4.
Gift of Seymour H. Knox Jr.

Figures 3.12 and 3.13 illustrate Hofmann's early use of color linked to an object's physical form.

FIGURE 3.14
Hans Hofmann, *Heraldic Call*, 1962. Oil on canvas, 153 × 122.9 cm (60¼ × 48⅜ in.). BAM/PFA, 1965.17. Gift of Hans Hofmann.

As the splashed paints of Hofmann's earlier work moved from foreground to background, the focus of his compositions shifted from external to internal relationships.

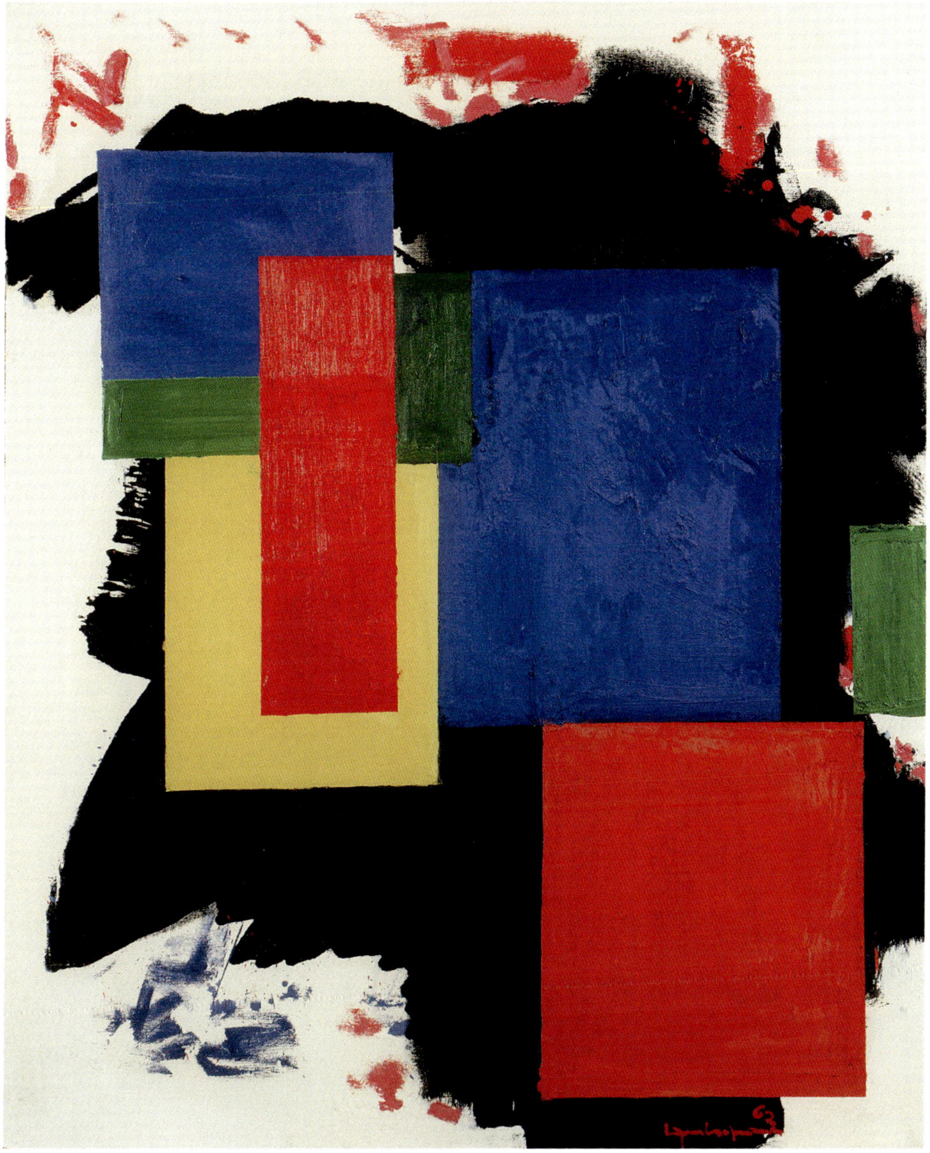

By the early 1960s Hofmann's focus had shifted from external to internal relationships. In the 1962 painting *Heraldic Call* (fig. 3.14), for example, the still-life environments of the artist's earlier work fall away as the viewer is pulled deeper into the composition. The hard-edge color planes of *Heraldic Call* do not interact with the black background form—they *emerge* from it. "The space in the object incorporates our objective world in its limits," wrote Hofmann, "and space in front of and behind the object, infinity" (1932–33, 4). Viewers are now positioned within the relational compositions they had previously observed from a distance. The relationships that define this painting are distilled from a more fundamental environment than those of the artist's earlier compositions. Such works as *Heraldic Call* simultaneously move beyond the object and position that object at the heart of the painting. In a similar manner, these paintings advance beyond the object-based color and compositional theories that remain central to their execution. In a late 1950s article entitled "Hans Hofmann: Grand Old Rebel," Clement Greenberg blamed Hofmann's delayed adoption of a personal style on the artist's "Cubist trauma" (1959, 27), but by the early 1960s Greenberg recognized in the artist's work an evolution of Cubist principles. Greenberg's 1961 monograph on the then eighty-one-year-old artist includes the suggestion that "Hofmann's Cubism, while becoming more outspoken than ever before...began at the same time both to vindicate and transcend itself—as if to purposefully refute what I had already said about it" (37).

Then, in the last months of his life, Hofmann turned his gaze to the forces at play within energy itself. In 1965 Hofmann explored the building blocks of existence in a handful of powerful works such as *Struwel Peter* (see fig. 3.9). The culmination of Hofmann's search for an object's essence, this primordial composition depicts the burst of energy from which all other relationships originate. These late pictures complete the viewer's journey into the object—from the energy of the object's environs to the energetic structure of the object—as the viewer at last perceives the interactive energies upon which the object's existence is founded. These unique compositions simultaneously exist in and out of the picture plane; they are self-contained and in balance with all creation. "These teeming canvases," critic Jesse Murray noted, "become not only metaphors of nature but of the fullness of life in the universe" (1981, 106).

Hofmann's push-and-pull color theories are a metaphor for the essence of reality at the heart of Hofmann's painting. "We experience totality of space as the result of forces and counter-forces that make a vital, force-impelled dynamic space," Hofmann told a Dartmouth College audience in 1962. "These forces and counter-forces reveal each other in varied but precisely defined tensions, which leads me to the conviction that space itself is energy" (1963a, 182).[56] In a lecture at the Smithsonian Institution ten years after Hofmann's death, student Fritz Bultman divulged that Hofmann sometimes called his works "quantum paintings" (Goodman 1986, 89), using a physics term defined as the smallest physical entity necessary for interaction. The distillation of relationships to their primary forces is an apt profession for the artist who lived in Munich when German physicist Albert Einstein received the Nobel Prize for work that defined the building blocks of energy, one of many advances in science and psychology that influenced contemporary modern artists.[57] Like Einstein, Hofmann chose an essential act—creation—and harnessed it, explored it in the world around him, and turned it in on itself. He explained this distillation in an August 1948 note entitled "Motto," found among the artist's personal papers: "Art is not an aesthetical manifestation only. Spiritual profoundness is inward orientation....It is through empathy in the spirit of nature and empathy in the spirit of the [painting] medium that this...is realized. When the message is truly profound and great, the aesthetical communication on which it is based is truly simple and pure" (Hofmann papers, [ca. 1904]–78).

Critical interest in the thematic and compositional consistency in Hofmann's late work arose shortly after his death. "These works...remain as dense and complex as when we first saw them," noted Metropolitan Museum of Art curator Henry Geldzahler in his catalogue for an exhibition of Hofmann's late paintings, "but they are beginning to look more and more like Hofmann's....It is now easier to recognize the qualities of style and personality that bind the individual works together" (1972, 10). "If the terms in which those [early and late] styles are presented differ slightly," historian Charles Millard later concurred, "the problems to which they address themselves are roughly the same" (1977, 404). The early roots of Hofmann's work have also begun to draw attention, as art historians such as Barbara Rose ascribe the power and directness of Hofmann's late-career canvases to his early lessons in the modernist distillation of color and form. "Hofmann achieved his own personal style not through a sudden existential epiphany," Rose noted (1978, 110), "but through a lifetime of laborious, painstaking progress toward a synthesis of the modernist tradition he had encountered along the way." This lifelong effort was also acknowledged by Kootz, who promised Hofmann lifetime representation and closed his gallery one year after the artist's death. In an accompanying article in *Arts Magazine* entitled "The Credibility of Color," Kootz reflected on Hofmann's journey: "I was happy for Hans, these past ten years. He was creating master work after master work, among which were the squares and rectangles, those 'irrational' areas of brilliance in color and final placements in space. 'These,' said Hans, 'are not easy. I work hard on them'" (1967, 38).

Hofmann's Materials as a Feature of His Late-Career Paintings

The success of Hofmann's most-metaphysical compositions is that these works simultaneously engage the viewer on a physical level. Hofmann's bold pigmentation and extremes of technique—what *ARTnews* editor Thomas Hess once referred to as "making a painting with almost nothing [to] making a painting with almost everything" (1965, 54)—emphasize the material construction of his artwork. As Greenberg noted in his 1961 monograph on the artist, "the weight and density of [Hofmann's] paint—attributes it has even when it is not thickly impastoed—contribute to the presence his pictures have as *objects* as well as pictures" (1961, 31). The transcendent message of Hofmann's canvases was achieved through ultimately corporeal means. "In those canvases that really 'come off,'" recalled Seitz, "the whole ensemble floats wonderfully before you: color, pigment, and the insistent geometry declaring themselves to the point where the surface is blasted even as it stresses the fact of its own physical existence" (1955, 113).

Hofmann saw no conflict in his simultaneous investigation of philosophy and materiality and applied his materials with the same enthusiasm he showed for the message they conveyed. "At the end of the day, he'd have [paint] all over his belly," recalled studio assistant Wolf Kahn, "like a palette himself" (Amgott 2003). "To explore the nature of the medium," said Hofmann, "is part of the understanding of nature, as well as part of the process of creation" (1948, 64, 71). In a typescript dated January 1956, he explained that making a painting was both "a physical and metaphysical performing" (Hofmann papers, [ca. 1904]–78).

Hofmann's paintings are a synthesis of Abstract Expressionist interest in the making and the meaning of art—a creative dialogue in which both the artist and his materials are active participants. Hofmann's work provides insight into not only the style but also the literal substance of this formative period of artistic innovation.

Existing Information Regarding Hofmann's Materials

Consistency in the construction and content of Hofmann's late paintings is matched by similarities in the accounts of Hofmann's practice and palette. Anecdotal accounts regarding Hofmann's studio practice and the limited number of scientific publications referencing his work provide an incomplete picture of the artist's materials but play an important role in the investigation of Hofmann's late-career paintings in allowing established beliefs regarding the artist's materials and techniques to be assessed in comparison with the materials analysis data provided in the present volume.

Documentation by Hofmann or others regarding the artist's materials is especially limited. Despite the prominent role of materials in Hofmann's compositions, the artist left little written documentation regarding his studio practices, and the specifics of his materials are rarely mentioned in reproductions of his writings or in interviews with him. Accounts of Hofmann's teaching and student notes from his lessons and lectures do not discuss the artist's own work—"Hofmann did not teach [about] himself," reported Glenn Wessels (Hofmann Student Dossier 1963). According to a November 1955 note page found among Hofmann's personal papers (Hofmann papers, [ca. 1904]–78), Hofmann felt that discussions that focused on technical matters took away from more important aspects of the final artwork:

> To the critic: Only the result counts
> not how it is done
> the means in themselves mean nothing
> as long as they expresses [*sic*]
> not something 'higher' in their relation
> they are the carriere [*sic*] only
> of the message.

The scant information provided by Hofmann regarding his materials is further reduced by editing trends in arts-related literature. The omission of detailed materials information was common in mid-twentieth-century publications, despite the important role of materials in modern art. "[Although] the meaning of art and its material form [were] inextricably linked," Carol Mancusi-Ungaro noted in her work on artist Jasper Johns, "editors...tend[ed] to skip over information about the minutiae of studio practice" (2007, 236). For example, the long-running *ARTnews* series that brought audiences into the studios of Abstract Expressionists including Hofmann and Willem de Kooning—a series dedicated specifically to artists' practice—provided generalized or exaggerated accounts of the artists' working methods or materials. The difference between the actual and reported practices of de Kooning, for example, has been well documented by Susan Lake, formerly the chief conservator and director of collections management at the Smithsonian Institution's Hirshhorn Museum and Sculpture Garden. "Despite the extent to which the general character of de Kooning's methods and materials have been broadly described," Lake noted, "writers confuse de Kooning's actual practices and tend to repeat [*ARTnews* editor Thomas] Hess's observations without confirming them through technical examination" (1999, 11–12).

The era's dismissive attitude toward technical documentation extended to scholarly archives. According to Janet Marontate, "The records [federal projects] saved were selected to provide information on arts administration policies and...political history, not to document technical or substantive aspects of the art produced" (1996, 74). Among the surprising practices catalogued by Marontate is the selective archiving of period technical records by the very government agencies established to compile and disseminate information about art materials:

> A "special skills" division was established in Washington to deal with practical questions confronting artists and craftsmen. This became the Handicraft Laboratory in Washington, DC and was renamed the Technical Services Laboratory in 1938....[One] function of the Laboratory was to review all technical instruction bulletins prepared or planned by state programs....In one month alone a report states that 47 technical bulletins sent by the Education Division were reviewed, yet *not one* of these or other technical bulletins mentioned in the correspondence has been found in the archives of the Laboratory. (Marontate 1996, 92)

The accounts of Hofmann's former students, studio assistants, and colleagues have likewise been marginalized or conflated with other documentation in subsequent publications about the artist. Artist and teacher Sam Feinstein, for example, when he studied with Hofmann from 1949 through 1952, filmed the artist at work, but the information he obtained regarding Hofmann's materials was edited from both the final film and the

FIGURE 3.15
Hofmann in his 8th Street studio, New York, 1960. Gloria Schoffel (later McDarrah) at left.

Photo: Fred W. McDarrah

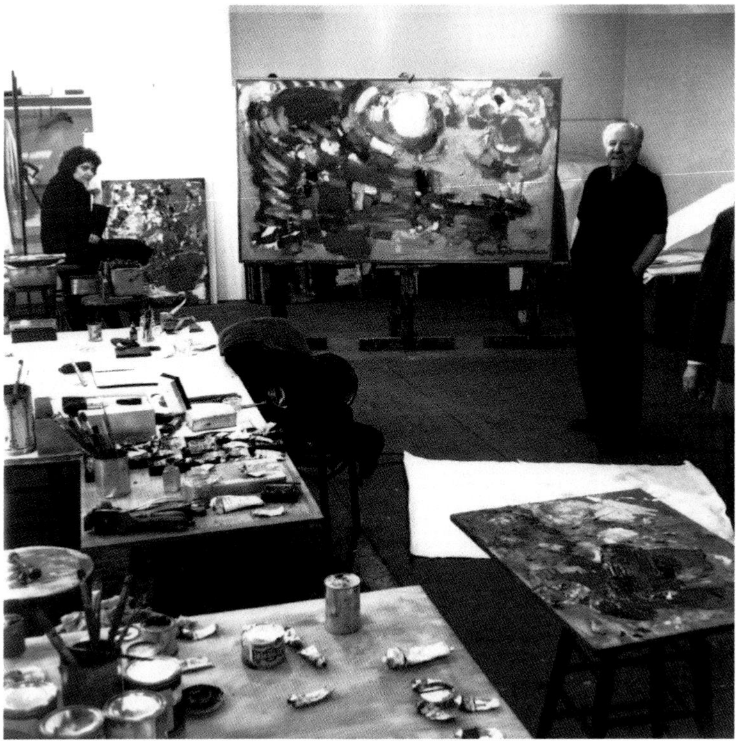

posthumous publication of an interview with Feinstein entitled *Portrait of Hans Hofmann as Painter, Teacher, and Friend* (2008).[58] Another discussion of Hofmann's studio practice—a chapter entitled "Notes on Technique" in Cynthia Goodman's 1986 monograph on the artist—mixes student recollections, information from a curator's essay recounting her visit to Hofmann's studio, and observations of conservators viewing the artist's work nearly twenty years after his death. Empirical and anecdotal information are not differentiated (the use of a conservator's report is mentioned only as a footnote), and no technical analysis accompanies the text.[59]

Hofmann was famously private about his studio practice (fig. 3.15). Hofmann student and American Abstract Artists cofounder Rosalind Bengelsdorf Browne said that Hofmann was so reticent to share information about his technique that some early members of the artists' group who studied with Hofmann did not even know that he was a practicing artist (Browne 1968). Although Hofmann held private meetings in his studio with colleagues including Franz Kline (Dickey 2011, 325), he did not welcome visitors when he himself was working. Feinstein—whose Provincetown studio was next door to Hofmann's—recalled that even in that collegial summer environment, students and colleagues had few opportunities to observe Hofmann at work: "I kept on pounding and finally [Hofmann] did come—I could hear the footsteps approaching—and he opened the door. He appeared before me like some dreadful apparition. He was covered in paint, as if he'd fallen into a vat with many colors.... His eyes were wild. He saw that it was me, and said, 'Ah, but at the moment I am not in the position to receive company. Perhaps you will come back'" (2008, 18). On the few occasions when Hofmann was observed, his behavior was influenced by the presence of an audience. "Hofmann himself was almost impossible to photograph," recalled Feinstein. "The phrase we use, of course, is 'self-conscious,' and what it really means is 'others-conscious.'... So if he was aware that he was being photographed, he would ham it up out of sheer nervousness and awareness of the filming" (2008, 23–24).

In 1950 two artists documented their observations of Hofmann at work. Elaine de Kooning made a series of visits to Hofmann's studio to watch the artist work on the painting *Fruit Bowl: Transubstantiation No. 1* and recounted the artist's progress in an *ARTnews* article entitled "Hans Hofmann Paints a Picture" (1950).[60] Feinstein subsequently filmed Hofmann at work on the painting *The Window*, for *Hans Hofmann: A Film by Sam Feinstein*, which Feinstein continued to edit until its first screening in 1999 at the Metropolitan Museum of Art.[61]

Despite focusing on the artist at work, neither account provides specific information about Hofmann's materials. "Because Hofmann was so well known as a teacher," recalled Feinstein, "the emphasis in the film is on his teaching principles even though it shows him making a painting" (Feinstein 2008, 28). De Kooning's article mentions only one specific painting material—the artists' oil paint Permalba[62]—and like Feinstein's, de Kooning's presentation of Hofmann is skewed toward ideological rather than practical matters. "Any discussion of Hofmann's technique," she wrote, "must revolve around his theories" (1950, 38). Furthermore, these accounts of Hofmann's technique are not consistent. For example, de Kooning reported that Hofmann said he "no longer has the patience to work with pen, pencil or crayon" (1950, 39), yet Feinstein's film shows Hofmann carefully plotting out his composition in pencil. Feinstein later said, "I was photographing it thinking: 'Well, I suppose I could blackmail him with this some day because who would believe Hofmann did such a thing?'" (2008, 21). Neither account may reflect the artist's true methods, as both accounts were susceptible to Hofmann's bouts of theatricality. "It was really difficult to get really decent photographs of Hofmann," concluded Feinstein, "unless he was totally unaware of what was happening" (2008, 24).

Other reports of Hofmann's methods may also contain elements of stagecraft. Kahn, Hofmann's studio assistant from 1947 through 1948, and Max Spoerri, studio assistant from 1956 through 1958, for example, have claimed that assistants stretched all of Hofmann's canvases (Kahn 2005; Spoerri 2005), yet when curator Katherine Kuh made a planned visit in 1958, she was greeted by the artist stretching his own supports: "To my amazement, I found Hans stretching canvases, most of them five to eight feet. There he was, kneeling, panting, hauling and with craftsman-like concentration turning out finished products of enviable precision. What astonished me was to see this artist then in his seventies expending energy on a physical chore that many younger men delegate to others" (2006, 257). "I always stretch my own canvases, glue them, and prepare them in every way," Hofmann later told Kuh. "It takes time but I simply cannot paint on commercial canvases . . . I've been working on the canvas for a long time before I start to paint it" (Kuh 1962, 128).

The available documentation regarding Hofmann's late-career materials is presented below. Firsthand documentation has been collected from Hofmann's publications, interviews, and personal papers. All references to materials in the artist's personal notes are in his handwriting unless otherwise noted. Other sources of information include interviews with former students and colleagues, unedited transcripts of previous interviews, and published articles and essays. Information gleaned from published analysis, empirical observations in collection surveys, and conservation treatment reports is discussed in chapter 4. The accounts of Hofmann's materials provided below are primarily drawn from the contemporary observations of Feinstein (published and unpublished) and Elaine de Kooning; from interviews conducted by this author, between 2002 and 2013, with Hofmann's former students and colleagues; and from the transcripts of similar interviews conducted by Goodman and Dickey and cited in their publications or housed at the Archives of American Art, Smithsonian Institution.[63]

Fabric Supports and Preparatory Materials

Contemporary accounts of Hofmann's late-career paintings mention the artist's use of both linen and cotton fabric supports. Spoerri (2005) said that Hofmann used "Belgian linen," which Erle Loran recalled to conservator Carolyn Tallent as "Utrecht linen" (1987), a likely conflation of Belgian canvas with the canvas supplier Feinstein referred to as "Utrecht linens,"[64] a Brooklyn-based importer later known as Utrecht Art Supplies. Conversely, de Kooning's 1950 article notes the artist's contemporaneous use of softer, cotton fabrics. "[Hofmann] usually paints on heavy [cotton] duck," she said, "originally for the sake of economy, but now because he finds it holds up better than linen" (40). Kahn also recalled Hofmann's use of cotton fabrics (2005), and the term "Heavy Duck" appears among the artist's papers on a loose, undated note page, as well as on a page within a journal marked "Provincetown 1951" (Hofmann papers, [ca. 1904]–78). The loose note page includes the notation "Rosenth." in a likely reference to Rosenthal's Art Supply at 47 East Ninth Street.[65] No information regarding the artist's use of cotton or linen fabrics was available from the local art stores frequented by Hofmann. Lillian Kiesler (Hofmann's school administrator, 1932–47) said that Hofmann bought prepared canvases from Lou Rosenthal at his eponymous art store located near Hofmann's studio at 53 East Ninth Street (1983, 3).[66] Kahn said that Hofmann purchased only rolled, unprimed fabrics and obtained materials from both Rosenthal and from Joseph Torch's store at 148 Fourteenth Street, above which Miz Hofmann, the artist's wife, maintained an apartment (Kahn 2005; Dickey 2011, 195). No archived records were available from either store. Rosenthal, who handled materials for Hofmann's students studying under the G.I. Bill of Rights and who stretched canvases for many local artists, did not archive his records, and in a 2005 interview he could not recall if he prepared any canvases for Hofmann (Rosenthal 2005). A sampling of Hofmann's tax records for the 1948–49 school year lists stretcher sets and rolls of fabric under the heading "Demonstration Materials," and an inventory of the "Main Studio & School" following the artist's death includes "1 large roll canvas 10' wide (Approx. 1000 sq. ft.)" and "several hundred sq. feet of canvas in pieces" (Hofmann papers, [ca. 1904]–78, "Estate Inventory," 10), but neither reference identifies the material as linen or cotton, and fabric rolls and remnants were not retained by the Hofmann Trust.

Both cotton and linen fabrics were discovered in the analysis of the study group paintings presented in chapter 5. Fabric supports for the study group paintings produced in the late 1940s exhibit the soft surfaces and bright fibers associated with cotton materials, while the fabric supports for study group works produced from 1953 onward exhibit the stiff, friable, and discolored characteristics common to aged linen canvases. Cotton fabrics may have been too flexible for the larger paintings produced by Hofmann in these later years, although the appearance of linen fabrics also coincides with Hofmann's growing financial success (and his ability to afford the more expensive linen).

Contemporary accounts of Hofmann's painting practices reflect some late-career shifts from traditional to modern methods and materials, primarily in the artist's preparatory layers. Traditional painting preparation includes "sizing" a raw canvas with glue to stiffen and seal the fabric fibers, followed by the application of a ground layer material to create a smooth painting surface and further isolate sensitive and absorbent fabric fibers from applications of acidic paint and varnish materials. Hofmann's 1930s studio practice was reported by Kiesler to be very traditional, including the use of aqueous glue and chalk-based gesso (1983, 3) to prepare the painting surfaces of his plywood supports. According to Kahn, this traditional ground layer treatment disappeared from Hofmann's practice with the artist's return to canvas in the late 1940s: "No gesso," Kahn said. "I never saw gesso in his studio" (2005). Gesso is not mentioned in the artist's 1948–49 tax return, although the undated "Rosenth." note includes a notation

FIGURE 3.16
Hofmann's New York Studio, 1950.

Photo: Rudolph Burckhardt for *ARTnews*.

for "Chesso," likely a generic (and misspelled) reference to the preparatory material.[67] "[Hofmann] prepares the raw canvas himself with flat white to close the pores," de Kooning reported in her *ARTnews* article, "then a gesso ground, which he maintains is the only ground that does not turn yellow" (1950, 38). In this case the term "gesso" likely refers to zinc white oil paint, a favored replacement for traditional lead white oil grounds that contained toxic material and often discolored with age, but should not, strictly speaking, be called gesso. Spoerri, for example, recalled Hofmann using "zinc white oil over rabbit skin glue [used to size the canvas]" to prepare his fabrics for painting (2005). The casual use of language in de Kooning's account—possibly loose terminology used by Hofmann during de Kooning's visit—confuses the vocabulary used to delineate the preparatory layers of a painting. The base layer of paint applied to a stretched fabric is properly called a ground layer, while subsequent overall layers of white paint are referred to as priming.[68] These terms are conflated in de Kooning's account of the artist's initial (ground) application of flat white followed by an overall application (priming) of "gesso" (likely zinc white oil).

The "flat white" ground layer mentioned in de Kooning's article may have been one of the oil-based alkyd house paints then popular with Hofmann's Abstract Expressionist colleagues (fig. 3.16). These alkyd paints did not stray far from the artists' familiar oil painting materials. The commercial alkyd house paint used by mid-twentieth-century artists was a long-oil mixture (more than 60% drying oil) of oil painting medium and

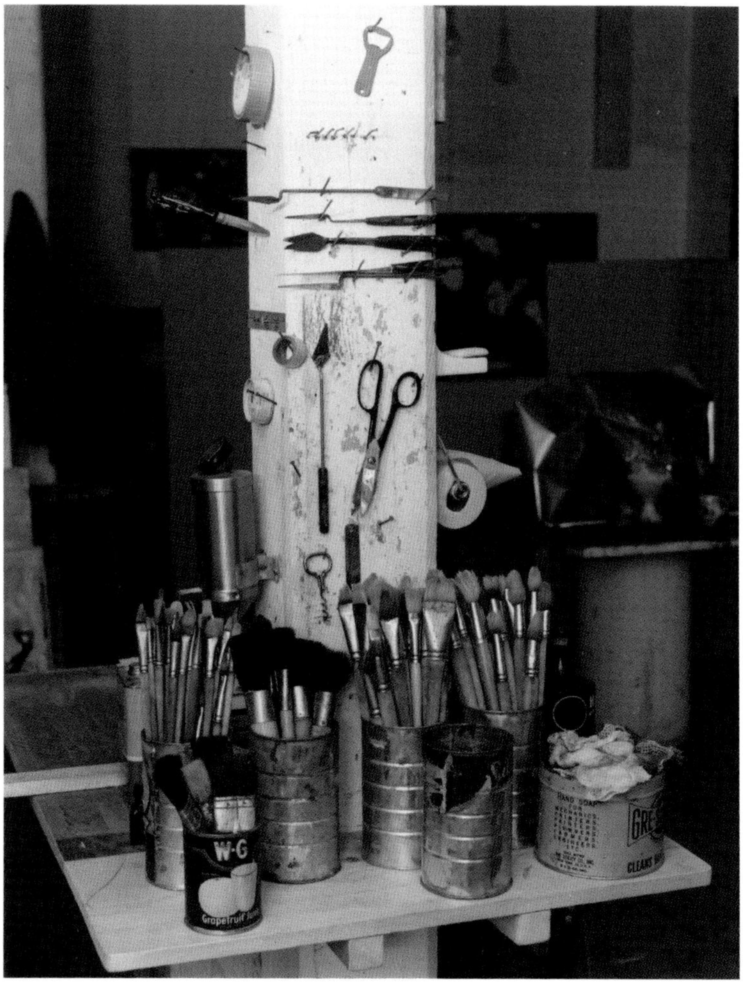

synthetic polymers that cured in the same manner as did oil paint, but at increased speed (Learner 2000, 99). Reports of Hofmann's use of house paint grounds appear in such unpublished accounts as "Conversation with Earl [*sic*] Loran, originator of the UC Berkeley Art Dept., and student of Hans Hofmann," dated June 9, 1981 (BAM/PFA records 1966–present), and Sascha Feinstein's 1989 interview of his father:

> When Utrecht Linens, which is now a big artists' supply house, were really starting in business, they sent two young men who came around to Provincetown to sell artists' materials. They came to me and said "Listen, we know that you're Hofmann's friend. Won't you tell him that he's doing the wrong thing to his canvas?" Hofmann, you see, used to buy raw canvas, and then would paint it white with oil house paint but without...any other preparation underneath it. This, naturally, was a bad thing to do to the fibers of the canvas. (Feinstein 1989)

Only one house paint is mentioned by name in the artist's personal papers, which contain a 1958 brochure for Benjamin Moore and Co.'s Alkyd Sani-Flat paint and a loose, handwritten note that reads, "Alkyd Santa [*sic*] Flat / White 20401 / Benjamin Moore & Co." A reference to "flat white" paint also appears in the only detailed statement by Hofmann regarding his materials. Typed and dated "Spring 1950" and presumably created in response to a curatorial inquiry, the statement parallels that year's *ARTnews* account of Hofmann's methods (de Kooning 1950). In the statement, the artist explains that "*Magenta and Blue* is painted in Rembrandt oil on heavy [cotton] duck on a gesso ground [likely zinc white oil priming] prepared by myself with flat white as underpainting [ground]" (Hofmann papers, [ca. 1904]–78). *Magenta and Blue*[69] was produced the same year as the *ARTnews* article, and both accounts note Hofmann's application of both ground and priming layers. Spoerri likewise said (2005) that it was not unusual for Hofmann to use a priming layer over an inadequately bright ground layer or to reground a discarded painting before beginning a new composition.

Compositional Paints

Contemporary accounts of Hofmann's compositional materials describe an exclusively oil-based palette. De Kooning notes Hofmann's use of water-based "tempera" paints for the paper-based preparatory sketches preceding his work on *Fruit Bowl: Transubstantiation No. 1* but mentions only artists' oil paints in descriptions of Hofmann's work on the actual painting (1950, 39).[70] Aside from Hofmann's alkyd ground materials, Feinstein's accounts of the artist likewise mention only oil paints (2008). A rare description of the paint colors and brands utilized by Hofmann is provided by Kiesler, who posed for the artist in the 1940s: "I can remember [that] I looked at him with something that amounted to awe and even a kind of greed at the number of boxes which he had—the cobalt violet, cadmium yellows, many, many cadmium yellows, and I can still see the labels on the Winsor [and] Newton and the Rembrandt. And the cadmium reds, and all the cadmium colors, and all the cobalt colors—large numbers of boxes.... He turned to me and said: '[With] color, Lillian, one is never stingy'" (1983, 1–2).

The colors and brands mentioned by Kiesler are in keeping with the artist's own records, including the earliest extant first-person documentation of Hofmann's materials—a handwritten list of paints in a combination of German, French, and English that reflects Hofmann's early artistic influences:

> Weiß [white]
> Perm.[alba] Weiß [brand name not legible]
> Green light ([F.] Weber [and Co.])

Verte Compose [mixed green] (Blockx)
Verte
Emeraude
Viridian (Rembrandt)
Cobalt [cobalt blue] (Rembrandt)
Light ultramarine/French ultramarine (Rembrandt)
Preuß[isch] Blaü [Prussian blue] (question spelling)
Schwarz [black]
Yellow (Blockx)
Cadmium yellow pale (Rembrandt)
Yellow light (F. Weber [and] Co.)
Yellow Mittel [medium]
Ocher
Cad[mium] orange [adjective not legible] (Winsor & Newton)
Cadmium rot deep [cadmium red deep] (Le Franc)
Vermilion
Yellow ocher
Laque de Garance [madder lake; brand name not legible]
Kobalt violette (Rembrandt)[71]

The list features colors and brands favored by Hofmann's German and French modernist colleagues. Colors on the list introduced shortly before or during Hofmann's lifetime include Permalba, the range of cadmium colors, cerulean blue, cobalt violet, and alizarin crimson (Boxall 1978a; Ball 2001a).[72] "Weiß" is likely a reference to zinc white, as Hofmann's refusal to use lead white was well known. "He used to tell me about all the artists in Paris who lived and worked in the same area," recalled Kiesler (1983, 2), "who got poisoning by using lead white. I can't remember all the ones that he named, but he had a certain fixation about being poisoned." The reference to "Perm. Weiß" may refer to Permalba, whose Philadelphia manufacturer, Weber, appears elsewhere on the list. Many of the same colors appear in a typed list of paint colors recommended to Hofmann's Provincetown students through the mid-1940s. The colors below are listed under the heading "Necessary materials for painting"; the colors marked with asterisks are noted as "essential":

* Permalba or zinc white
* Cadmium lemon, light, medium, orange, and red light
Vermilion
* Cerulean deep
* Cobalt blue deep
* Ultramarine blue deep
* Emerald green
Cobalt violet
* Alizarin crimson
* Yellow ochre
* Light red
* Burnt sienna
Mars violet
* Burnt umber
* Ivory black[73]

This list is leaner, with an added selection of neutral brown and black pigments likely intended for the painting exercises assigned to students following their transition from drawing to painting curricula. Cerulean blue and Mars violet are new additions that do not appear on the 1930s list. Alizarin crimson is the synthetic substitute for the previously listed madder lake; emerald green remains despite the existence of synthetic alternatives such as viridian green. European paint manufacturers Winsor and Newton, Rembrandt, and Blockx reappear (in a side note) as recommended brands for high-quality oil paints. "I remember the first time when I was there when he was starting the portrait," recalled Kiesler (1983, 2), "that he took one of those tubes, Winsor and Newton cobalt violet and it seemed to me in one squeeze he emptied the entire tube on that palette. I was aghast [at the expense]." European manufacturer Schmincke and American manufacturers Permanent Pigments and F. Weber and Company are also recommended as less expensive options. Loran said he thought Hofmann occasionally may have used Grumbacher oil paints, and according to Kahn, in at least one instance Hofmann utilized a set of Bellini brand oil paints made by manufacturer Leonard Bocour and given to Hofmann by the paint maker when he visited classes in Provincetown (Dickey 2011, 243; Kahn 2005; Tallent 1987).

Permalba appears on both lists and is one of the few paints regularly mentioned in both published and unpublished accounts of Hofmann's materials. Both Kiesler and Elaine de Kooning noted Hofmann's use of the popular white paint, a color Spoerri specified was used by the artist for compositional painting only (Kiesler 1983, 1–2; de Kooning 1950, 58; Spoerri 2005). "He [used] a mountain of Permalba," recalled Kiesler. "[He] squeeze[d] out what seemed to me to be half a tube or an entire tube of [it]" (1983, 2). Twenty-two tubes of Permalba are included in the artist's tax records for the 1948–49 school season (Hofmann papers, [ca. 1904]–78). In addition to the tubes of Permalba paint, Hofmann's account of oil paints used as "Consummable Instructional Supplies" during the 1948–49 school year included 120 tubes of oil paint valued at fifty cents each and 60 tubes of oil paint valued at seventy-five cents each. Ten tubes are noted as "Cad[mium] Color"; no other colors are specified. Trust records note "several hundred tubes of oils (still in good condition)" among the contents of the "Small Workroom of Studio" after the artist's death, but the records do not provide further identification of the materials (Hofmann papers, [ca. 1904]–78, "Estate Inventory," 10). None of the artist's paints are in current trust holdings, and no pigment lists remain from the 1950s or 1960s.

Solvents, Mediums, and Varnishes

Accounts by and about Hofmann include mention of the use of thinners and other additives to manipulate the handling characteristics of the artist's paints. In addition to general cleaning uses, solvents and diluents can be used to thin the consistency of the paint, while varnishes and extra paint medium can be added to paint in order to thicken it, increase the transparency of a color, or alter the surface sheen of the paint once it is dried. As more additives are used, less paint is needed. "[I] may use a hundred tubes for one picture," Hofmann told de Kooning, "or one tube for a hundred pictures; lots of medium or none at all" (de Kooning 1950, 40). Kiesler, Loran, and Spoerri all recall turpentine among the artist's painting supplies, and this common diluent also appears in the accounts of both Feinstein (2008, 22) and de Kooning: "[Hofmann] picked up one of the paint-soaked pieces of gauze that had accumulated on his palette table, wrung it out so that it was almost dry, dragged it across the coarse-grained duck for a dry-brush effect; dipped another piece into turpentine (the only medium used for this picture) and washed on a glaze" (1950, 58).

Other thinners and additives appear in Hofmann's records and reports of his studio practice. "Anything can serve as a medium," Hofmann joked, "kerosene, benzine,

FIGURE 3.17
Hofmann at work in his New York studio, 1950.

Photo: Rudolph Burckhardt for *ARTnews*.

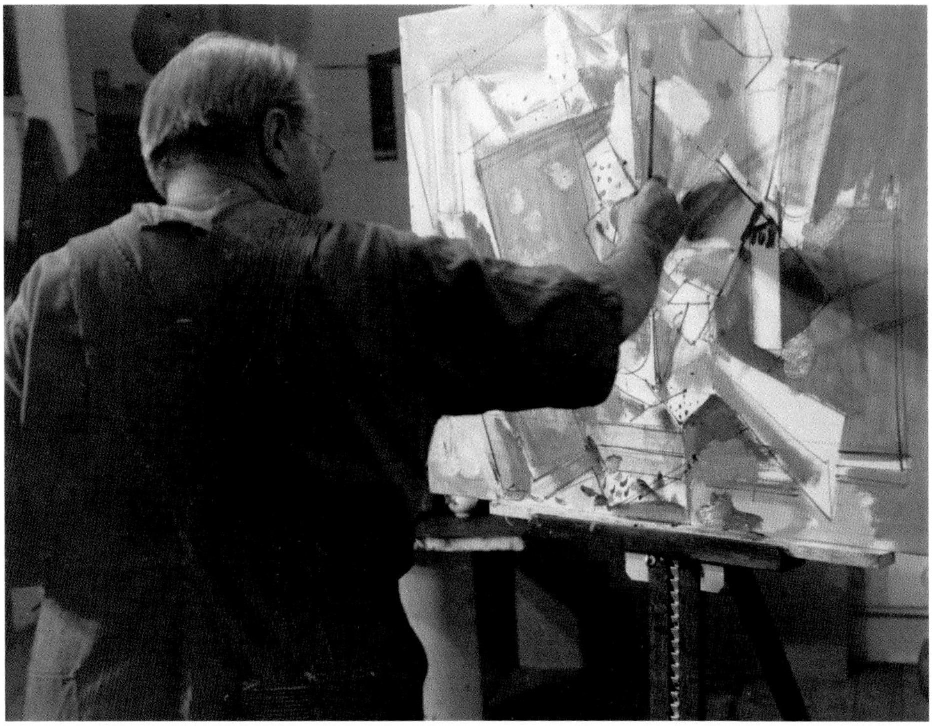

turpentine, linseed oil, beeswax...even beer" (de Kooning 1950, 40) (fig. 3.17).[74] Hofmann student Erik Koch recalled that a popular artist's medium in the late 1940s and early 1950s combined turpentine, linseed oil, and stand (thickened) oil or varnish (Koch 2012), a mixture mentioned in contemporary articles on the studio practice of Abstract Expressionists Jack Tworkov and John Ferren.[75] Turpentine, benzine (petroleum ether), and linseed oil appear in Hofmann's 1948–49 tax records, and both "rectified spirits of turpentine" and "purified linseed oil" are listed on the student supply list under the heading "Necessary materials for painting" (Hofmann papers, [ca. 1904]–78).

There are no accounts of varnish coatings in Hofmann's painting practice, and the artist made his opinion regarding varnish very clear in the *Magenta and Blue* statement when he asserted, "I never glaze or varnish a painting" (Hofmann papers, [ca. 1904]–78). Two documents in the artist's papers are in conflict with this statement. The estate inventory lists "4 large bottles retouching varnish" among the contents of the "Small Workroom of Studio," and there is a loose page among the artist's financial records that contains the following notations in an unknown hand: "methacrylate (final picture varnish—can be made mat or dull with addition of aluminum stearate)" and "varnish put out by Ralph Mayer (dull or mat)."[76] These references to synthetic varnish may be posthumous and unrelated to Hofmann's actual studio practice. The inventory of Hofmann's studio, for example, was not completed until six years after his death and may include materials used to varnish the works before distribution by the Trust.[77] The artist's papers were compiled during the same period and may therefore contain documents related to the posthumous care of Hofmann's paintings.[78] While this information may not relate directly to Hofmann's use of materials, it may be useful in distinguishing between original and nonoriginal material during the scientific analysis of Hofmann's paintings. A history of conservation varnishing of Hofmann's work is provided in chapter 4.

4 Selection and Conservation History of the Paintings in This Study

Painters must speak through paint—not through words.
—HANS HOFMANN, "Statement," *It Is*, 1959

Hofmann has ~~often~~ *frequently* remarked of his ~~media~~ *materials*, with a mixture of respect and mock irritation, that "they fight back."
—SAM HUNTER, draft of a 1964 essay, with editorial changes by Hofmann

Understanding Hofmann's material preferences and late-career shifts in palette requires the collection of a large amount of data. A large group of paintings representative of the range of materials noted in the available literature was therefore chosen for comprehensive examination and analysis. The study group was fashioned to reflect Hofmann's stylistic evolution and the condition history of his paintings in order to reveal any relationship between the artist's materials and these aspects of his late work.

Selection of the Study Group Paintings

The bulk of the late-career paintings chosen for this study came from the Hans Hofmann Collection at the University of California, Berkeley Art Museum and Pacific Film Archive. The Berkeley collection is the world's largest public collection of Hofmann paintings and contains a large and representative sampling of the artist's late-career work, accompanied by lengthy archival records of the paintings' condition and conservation treatment histories. The university's Hofmann Collection was established by a direct gift from the artist. Three years before his death, Hofmann promised forty-seven of his paintings, along with a donation of a quarter of a million dollars, to the University of California, Berkeley, in recognition of the university's role in helping Hofmann escape the deteriorating situation in Germany and build a new life in the United States. "If I had not been rescued by America," Hofmann confided to art historian and curator Peter Selz, "I would have lost my chance as a painter" (Selz 2002, 5).[79] Hofmann's intentions were first communicated in a July 25, 1963, letter from former Hofmann student and University of California, Berkeley, art department faculty member Erle Loran to university president Clark Kerr: "The magnificent generosity of Mr. Hofmann toward the University at Berkeley can only be understood when he tells you in his own words of the momentous change that occurred in his life when the late Professor Worth Ryder recommended that [Hofmann] should be invited to teach at Berkeley in 1930. He is convinced that he would have been eliminated by the Nazi government had he stayed in Germany" (BAM/PFA records 1966–present, Loran to Kerr, July 25, 1963, 4).

Hofmann's gift was also intended to jump-start the university's fund-raising for a purpose-built art museum to house its growing collection. The monetary portion of Hofmann's gift was granted through partial interest in a selection of the artist's works to be sold through dealer Samuel Kootz, who helped select the paintings to be gifted to the university. The initial selection of works for the Hofmann Collection coincided with a retrospective of the artist's work at the Museum of Modern Art organized by curator William Seitz, who allowed Hofmann, Loran, and Kootz to walk the exhibition galleries and select paintings for donation and sale.[80] "Both Kootz and Mr. Hofmann suggested that the three of us work together in choosing the best paintings for the University before the exhibition opens to the public," Loran explained to Kerr in his letter. "Mr. William Seitz, the painting curator of the Museum of Modern Art, who is responsible for organizing and writing the catalogue for the Hans Hofmann exhibition, has very graciously offered to make the galleries available to us for this purpose" (BAM/PFA records 1966–present, Loran to Kerr, July 25, 1963, 2). Ownership of the first group of paintings transferred to the university on December 27, 1963. Ownership of additional paintings was transferred to the museum on March 5, 1964, and on January 20 and May 21, 1965. In 1965 Selz left his curatorial position at the Museum of Modern Art to become founding director of the university's new museum,[81] and he assisted in the final selection of paintings from Hofmann's studio; ownership of this last group of works transferred to the university on January 26, 1966, just weeks before Hofmann's death (Selz 1993, 127).

Twenty-six paintings and eight palettes were chosen for study and analysis as exemplars of Hofmann's late-career work. Gaps in the Berkeley collection were filled by selected additions from the collections of the Albright-Knox Art Gallery in Buffalo, New York; the Memorial Art Gallery, Rochester, New York; and the Museum of Modern Art, New York. The paintings selected represent the range of late-career materials and paint application methods noted in the available literature. Each variation in the artist's late style is represented by at least one painting in the study group, and by two examples when possible. The study group includes twenty-three paintings on canvas from the years 1953 through 1965, a period that includes the years just prior to and after the 1958 closing of Hofmann's schools in New York and Provincetown, Massachusetts. Three works on canvas from the late 1940s were also included so that any change in materials between these earlier, experimental works and the artist's more established late-career style could be identified. Paintings on panel were eliminated from consideration because reactive plywood panel supports would impede the observation of paint-related condition issues. Eight palettes on various supports found in Hofmann's studio at the time of his death were included in order that any late experimentation with new materials could be noted.

Figure 4.1 is a time line spread of the study group. The first half of the time line includes the years when Hofmann was at the height of his popularity as a teacher: "at any given time in the last year[s]," recalled former student James Gahagan, "there were a minimum of seventy-five to a hundred students working in the studio—roughly twenty-five and thirty in each session....[and] about 250 students...each summer" (Bultman 1978); the second half of the time line includes Hofmann's most prolific and recognizable period of artistic production. Table 1 (see appendix 1) presents a chronological listing of the study group paintings.

FIGURE 4.1
Time line spread of the study group paintings; palettes found in Hofmann's studio at the time of his death have been dated 1966. The red line indicates the year Hofmann closed his schools and began to paint full-time.

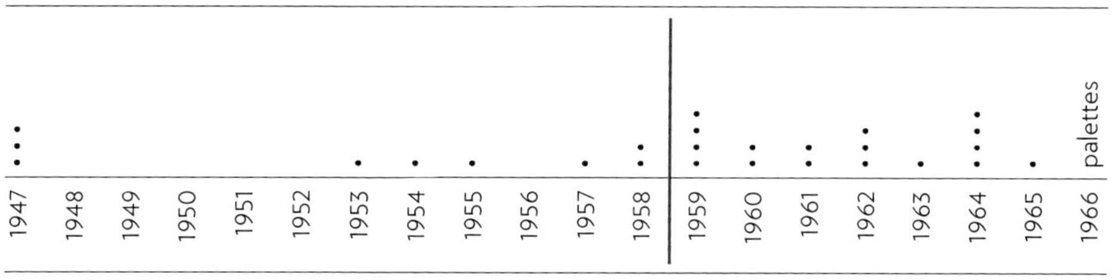

Physical History of the Study Group Paintings

The study of Abstract Expressionist materials not only enhances our understanding of modern art but also assists in the preservation of its legacy. The work of the Abstract Expressionists set as standard modernist practice a combination of experimental materials and techniques or, according to former Mark Rothko Foundation conservator Dana Cranmer, the "unpredictable formulations of the...medium, and the unconventional techniques [of the artist]" (1987b, 196). Key works from this period are fragile, susceptible to accelerated aging, and difficult to treat. "If one is conservator to a museum that...adds contemporary art to its permanent collection," said Margaret Watherston, former paintings conservator for the Whitney Museum of American Art, "one must be ready for the problems that are likely to occur with these acquisitions" (1974, 119). Tracking the appearance of new and modified paint materials in the work of Abstract Expressionist painters and investigating the artists' use of these materials are endeavors that add much-needed documentation to the growing catalogue of twentieth-century art studies, assist scientific studies currently under way regarding the long-term behavior of these materials, and directly impact the future conservation treatment of Abstract Expressionist and later works that incorporate unconventional materials into an otherwise traditional artist's palette. "If we fail to properly research our [modern] collections," warned Stedelijk Museum director Rudi Fuchs, "we neglect our cultural heritage" (Hummelen et al. 1999, 312).

A brief discussion of the Abstract Expressionists' relationship with modern paint materials may help put Hofmann's use of these materials in context. Many mid-twentieth-century artists were experimenting with changeable materials, products in the early stages of commercial production, or products designed for the transitory needs of a mass market. Zinc oxide oil paints, for example, were a favorite choice for Abstract Expressionist ground layers (Rogala et al. 2010) during a period of experimental zinc oxide formulation that produced stiff and brittle commercial paints that were ill-suited for supporting heavy Abstract Expressionist compositions (Rogala et al. 2010; Maines et al. 2011; Osmond 2014). "Many of the materials employed in the making of a contemporary painting are known to have been designed for a very limited useful life sufficient for a special need," cautioned Louis Pomerantz at a 1962 Artists Equity meeting in Chicago—"not always for [the need] of the artist, whose work is intended for greater permanence" (1962, 54). Other vagaries of early industrial paint formulation include such World War II–era substitutions as formaldehyde biocides and postwar surplus styrene-butadiene rubber plasticizers,[82] or shifts in formulation resulting from wartime innovations, including changes in such familiar industrial paints as Duco, originally a cellulose nitrate–based formula that later incorporated a polyester-modified oil medium that was a specialty of the Delaware-based explosives maker turned paint maker E.I. du Pont de Nemours and Company.[83]

Condition issues related to experimental paint formulations can be compounded in Abstract Expressionist works by the artists' unconventional manipulation of their materials.

Abstract Expressionist compositions that push the boundaries of paint application likewise test the physical limits of their material components—for example, in vibrant colors and subtle surfaces achieved through the extreme thinning of paints. While modern paint manufacturers strove to avoid pigment-binder separation (Golden 2003), a new approach to color effects was achieved by Abstract Expressionist painters through direct manipulation of manufacturer formulations. According to reports by critic Clement Greenberg and others, the artists' groundbreaking stain painting technique was achieved by "dilut[ing] paint to an extreme and soak[ing] it into unsized and unprimed canvas" (Greenberg 1964, 14). Cranmer noted how Rothko's use of solvent-thinned paint "ignored the limits of physical coherence to achieve a translucency" (1987b, 192) unique to thin layers of underbound pigment, which allowed light "to penetrate directly, its passage unencumbered by medium, to strike the individual pigment particle and bounce back, affording the viewer a rare experience of clear, resonant color" (Cranmer 1987a, 285). The resulting surfaces are brittle and easily abraded, and they leave underbound pigments vulnerable to environmental exposure.

The surface vulnerabilities of thinned paint films are also seen in works by Abstract Expressionists who encouraged pigment-binder separation in order to obtain heavily pigmented paint films. Solomon R. Guggenheim Museum deputy director and chief conservator Carol Stringari noted that in Ad Reinhardt's studio, paint cans "would be left undisturbed, allowing the pigments to settle to the bottom of the jar, leaving the oil/turpentine extract above the dense pigment phase. This oil would then be poured off, with only enough retained to allow the paint to be brushable." This technique gave the artist's paintings "a beautiful...suede-like surface, similar in appearance to pastel" (Stringari, Pratt, and McGlinchey 2004, 165), but like the underbound surfaces of work by his colleagues Helen Frankenthaler and Rothko, Reinhardt's delicately balanced surfaces are vulnerable to abrasion and easily absorb oils and dirt from fingerprints and atmospheric pollutants. Reinhardt once noted, "the picture leaves the studio as a purist, abstract, non-objective work of art [but] returns as a record of everyday (*surrealist, expressionist*) experience ('chance' spots, defacements, hand-markings, accidents, 'happenings,' scratches)" (Lippard and Hunter 1967, 22).

On the opposite end of the spectrum of paint application are Abstract Expressionist compositions constructed with heavily applied layers of paint. The work of such Abstract Expressionists as Clyfford Still and Hofmann regularly pushed the limits of canvas and ground layer supports by alternating open expanses of canvas with heavy paint layers in compositions described by Seitz as "run[ning] the gamut from the uncovered canvas to a reckless loading, which forms a veritable cliff of pigment" (1955, 21). Many of Hofmann's signature works are composed of multiple heavy paint layers, described as "sculptures-in-paint" by Frederick Wight. "[Hofmann] almost literally weights the surfaces," wrote Wight, "and seems to be composing in depth by means of the differing specific gravities of the colors on his palette" (1957, 44).

Some of Hofmann's colleagues added their own materials to commercial formulations. Jackson Pollock and Willem de Kooning, for example, added everything from studio trash to seawater to their compositions. "Less recognized is the extent to which [de Kooning] consciously employed unconventional materials in his paintings," Lake noted (Lake, Lomax, and Schilling 1999, 381).[84] Rothko not only thinned solvent-sensitive paints with water and turpentine but also experimented with a homemade egg-oil emulsion that Mancusi-Ungaro reported as "consist[ing] of unmeasured amounts of the tube oil paint, whole beaten eggs, turpentine and dammar resin" (1990, 135). Early formulations of acrylic paints utilized varying acrylate:methacrylate ratios unrelated to previous solution or dispersion formulations, and such small-run manufacturers as New

York's Samuel Golden and Leonard Bocour of Bocour Artist Colors regularly changed their Magna acrylic paint formulations to address requests for different handling characteristics. Henry Levison of Cincinnati's Permanent Pigments frequently used artists to test new formulations (Marontate 1996, 223); Bocour and Golden produced old-recipe or custom formulations for Morris Louis (Gates et al. 2005, 330; 2007, 277); and Pollock is reported to have received custom batches of oil-based paint from commercial manufacturers (Rose 1980).[85] Shifts in formulation can produce different aging behaviors in works from the same period by a given artist, as well as in works made with paints from the same manufacturer. "Knowledge regarding how well any of these modern paint media will withstand the passing of time remains extremely limited," noted Thomas J. S. Learner, head of Science at the Getty Conservation Institute (GCI 2002, 18).

Materials Noted in Conservation Assessments and Research of Hofmann's Work

Limited information related to Hofmann's materials exists in the conservation and research literature. Empirical observation—but no technical analysis—of Hofmann's materials is available in two unpublished surveys performed in the 1980s of the Hofmann Collection at the University of California, Berkeley Art Museum and Pacific Film Archive. In 1982 San Francisco Museum of Modern Art conservator and former Hofmann student Thornton "Tony" Rockwell surveyed the treatment history and condition of the Hofmann Collection and compiled the information into an unpublished report entitled "Conservation Problems Present in Paintings by Hans Hofmann" (Rockwell 1982). Six years later conservator Carolyn Tallent performed a similar condition assessment of the Berkeley collection as part of her postgraduate fellowship at the Intermuseum Conservation Association in Oberlin, Ohio (now located in Cleveland), and produced an unpublished report entitled "Investigation of the Painting Materials and Techniques of Hans Hofmann—Preliminary Report" (Tallent 1988).[86] No scientific analysis was performed for either survey.[87] Previous scientific analysis of Hofmann's materials by this author includes limited scientific data related to the ground layer materials of a small number of late-career palettes (Rogala 2005)[88] and the examination of six late paintings during a study of Abstract Expressionist ground layer materials in the collection of the Hirshhorn Museum and Sculpture Garden of the Smithsonian Institution (Rogala et al. 2010; Maines et al. 2011).[89] No other analysis related to Hofmann's materials has been found in recent searches of published conservation and scientific literature.

Mentions in the conservation literature of preparatory materials related to Hofmann's paintings concur with the accounts of the artist's students and colleagues. Conservators Rockwell and Tallent both note a transition from plywood to fabric supports in the late 1940s and confirm Spoerri's account of glue sizing on Hofmann's canvases. Rockwell bases his identification of an isolating glue-size application on the lack of binding media bleed-through to absorbent canvases, while Tallent based her sizing identification on an observed ultraviolet-induced visible fluorescence consistent with similar protein-based sizing in paintings examined under ultraviolet illumination (Tallent 1987; 1988, 6). Hofmann's possible embrace of house paint ground materials is also mentioned in the conservation literature. Rockwell suggested that during the 1950s, "alkyd white interior house-paint replaced glue gesso [used on earlier plywood panels] primings [grounds] to become the principal ground used on Hofmann's canvases" (Rockwell 1982, 10, 14),[90] and Tallent noted that the thin, matte ground materials applied during this period conformed to the underlying support fabric in a manner consistent with alkyd paint.[91] Handwritten notes accompanying Tallent's research materials also state that she observed ultraviolet-induced visible fluorescence consistent with alkyd paints in Hofmann's work from the mid-1950s onward (1988, 7, chart). Analysis performed in this author's earlier research identified alkyd ground layers on two paperboard palettes from Hofmann's studio.[92] In that study, pyrolysis–gas chromatography–mass spectrometry (Py-GC-MS) and Fourier transform infrared spectroscopy (FTIR) were used to confirm a close match between the ground layers applied to the palettes and paint found under the lip of a paint can lid also used by Hofmann as a palette.[93] The can lid bore a series of embossed numbers identified by archivists at Benjamin Moore and Company as a match for a March 17, 1957, batch of Benjamin Moore Alkyd Sani-Flat 20401 Ultra White (Rogala 2005, 8). Brochures for this paint were found among Hofmann's studio papers (Hofmann papers, [ca. 1904]–78).

Limited information regarding Hofmann's compositional paints appears in the conservation literature. Notable exceptions are Rockwell's mention of a "black enamel 'stove' paint" and "poured black paint" in works from 1959 to 1962 (1982, 13-A), alizarin-like colors in works produced as early as 1947 and as late as 1965, and the late 1950s appearance of "magenta or rose-red lake pigments," an attribution Rockwell credits to Berkeley Art Museum and Pacific Film Archive archivist Lawrence Dinnean (1982, 14, 16). Rockwell's use of the term *stove* may be a generic reference to industrial alkyd paints of that period, many of which were dried under heat, or "stoved." Rockwell also refers to Hofmann's poured paints as "Duco" enamels, another often-used appellation for alkyd paints. Rockwell does not recall the source of these identifications (2012). The only other mention of alkyd compositional paints in the conservation literature is found in a loose note among Tallent's research papers regarding a single instance of alkyd priming materials observed in the mixed color of a single color plane.[94] A small number of Hofmann's pigments were analyzed in this author's previous research. In situ X-ray fluorescence (XRF), as well as sampling and subsequent analysis through optical microscopy and scanning electron microscopy–energy dispersive spectroscopy (SEM-EDS), were used to identify zinc oxide white, cadmium yellow and red, and cobalt blue in the five canvas works produced by Hofmann between 1957 and 1963.[95] The same pigments are among the paint colors identified in the study group works examined in this book.

Condition Issues in Hofmann's Late-Career Paintings

Hofmann's late-career paintings test both the aesthetic and physical limits of his materials. The paintings selected for this study were chosen to represent not only the artist's wide-ranging paint application methods but also the condition issues common to Hofmann's late works. Caretakers of Hofmann's work are faced with complex situations involving distorted canvas, lifting paint, and slow-drying or improperly drying paints, yet little information exists regarding the origins of these phenomena. When discussing Hofmann's work, it is also important to distinguish between visual evidence of the artist's technique—also known as facture—and those material-related behaviors that affect the overall stability of an artwork. A review of material-related behaviors in Hofmann's late-career paintings is presented below to aid in differentiating evidence of the artist's technique and aging behaviors common to works on canvas from those condition issues that may be directly tied to Hofmann's choice of materials. In order to understand the appearance of nonoriginal materials in Hofmann's work, an overview of conservation materials specifically noted or in common use during the preservation of the study group paintings is also provided.

Information regarding the condition and conservation of the study group paintings was compiled from this author's own observations during selection and sampling of the study group works as well as from the archived conservation records of the Albright-Knox Art Gallery, the Memorial Art Gallery, and the Museum of Modern Art, and particularly the conservation records related to the Hofmann Collection at the Berkeley Art Museum and Pacific Film Archive, including the ten condition surveys of the

Hofmann Collection undertaken between 1974 and 2009 by the conservation staff of the San Francisco Museum of Modern Art.[96] Physical phenomena observed or reported in each of the individual study paintings are compiled in table 2 (see appendix 1).

Fabric and Wooden Support Materials

The aging behaviors of traditional support materials can be incompatible with the heavy, sculpted surfaces of some Abstract Expressionist paintings. Cotton canvases stretch, and linen canvases became slack and brittle with time, and this distortion exaggerates the stress that heavy paint layers impose on absorbent or weak ground layer materials and wooden commercial stretchers, which are designed to maintain tension in only two dimensions. Hofmann's late-career paintings are physical as well as optical constructions that, according to conservators, test the limits of their support structures. "Although the stretchers [Hofmann] used were generally sound, with cross bars appropriate to size," noted Rockwell (1982, 9), "they were sometimes not sturdy enough to support [his] larger works with weighty paint films." Canvas distortion is advanced in Hofmann's paintings with absorbent grounds or unsized canvas materials that leach paint medium and place acidic materials in contact with canvas fibers. Despite assertions that Hofmann sized his canvases, references to unsized canvases do appear in Rockwell's records (1982, 9), and examination of the study group paintings has revealed purposefully uneven ground layer preparation. Distorted or unevenly weakened canvases exhibit uneven responses to environmental factors such as humidity, imposing further imbalanced stress on the painting structure. According to condition survey reports, eighteen of the twenty-six study group paintings have exhibited some level of canvas distortion, and examples of localized distortion appear in at least four works produced by Hofmann in the two years following his retirement from teaching (1959–60; see figs. 4.2 and 4.3).

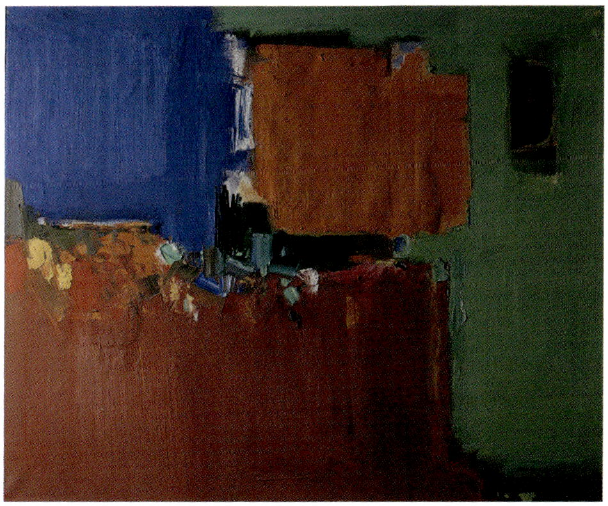

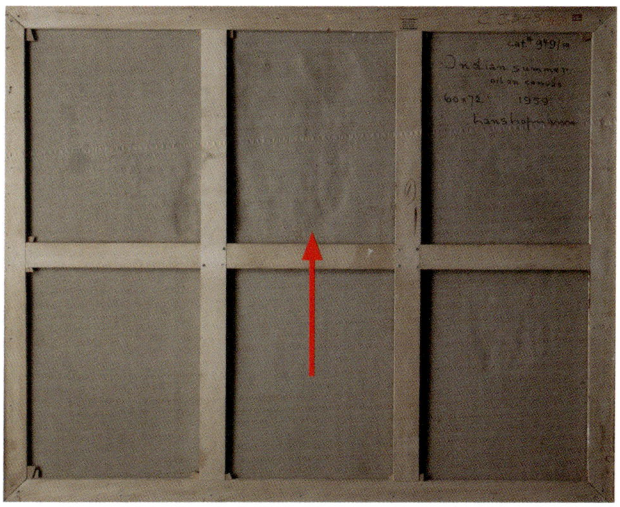

FIGURES 4.2 AND 4.3
Hans Hofmann's *Indian Summer* (1959; see fig. 3.7), photographed in oblique light.

These images illustrate regional distortions that appear in some of Hofmann's late-career paintings.

Paint and Preparatory Surfaces

Supporting and compositional paint layers can exhibit a variety of behaviors related to application technique or to the artist's choice of materials. Stress can originate from within the painting or be applied by an external source. Stress imposed on a painting before, during, or after application of a paint layer, for example, can lead to breakage within the paint film (intralayer cleavage) or failure of the bond between the paint film and neighboring materials (interlayer cleavage). Cleavage in the paint film may remain stable under additional stress, or it may transfer new stresses to neighboring materials in a chain reaction that creates material failure throughout the painting. Weak or brittle paint films can fail under stress; tough or flexible materials may transfer their stress to weaker

FIGURES 4.4 AND 4.5
Details of Hans Hofmann's *Struwel Peter* (1965; see fig. 3.9) and *The Clash* (1964; see fig. 3.11), photographed with normal illumination. The violet paint film on the left exhibits traction cracks related to oxidative drying, as does the thicker paint film on the right, where cracks have developed deep within the film and appear as depressed lines in the paint surface. The paint film detailed on the right also exhibits a surface wrinkling phenomenon.

FIGURES 4.6 AND 4.7
Details of Hans Hofmann's *The Third Hand* (1947) and *Scintillating Space* (1954), photographed with normal illumination. The brittle white paint film on *The Third Hand* (left) exhibits widespread microcracking in response to stress, while stress imposed on the equally thick but tougher green paint film on *Scintillating Space* (right) results in deeper, less frequent cracks.

paint films. A range of paint and ground layer failures commonly appears as paintings age, but failures do not necessarily lead to paint loss or endanger the overall stability of the artwork. Conservation records indicate that all of the paintings examined in this study have exhibited some level of paint cracking or cleavage.

Oil paint films are stressed during the cycles of expansion and contraction associated with oxidative drying. Some paints exhibit contraction, or "traction," cracks in response to this stress. Traction cracks in thin paint films can separate and expose the underlying ground layer (fig. 4.4), while cracks in thicker paints may develop deep within the film and appear as depressed lines in the paint surface (fig. 4.5). Brittle films may show widespread cracking (fig. 4.6), while tougher paint films exhibit less frequent, deeper cracks (fig. 4.7).

While the appearance of cracks is not a condition problem in itself, condition issues related to cracking paint do appear in Hofmann's late work. Some of these crack-related concerns were evident even during the artist's lifetime. On January 24, 1964, the Museum of Modern Art exhibitions director wrote to the associate counsel for the Regents of the University of California to express his concerns regarding *Tormented Bull* (1961; see fig. 3.8), then traveling in Hofmann's retrospective exhibition and destined for donation to Berkeley: "As you will see from the attached copy of a letter from Mr. Hofmann," the letter reads, "that painting has an inherent flaking condition whose further deterioration cannot be covered by insurance" (BAM/PFA records 1966–present, Waldo Rasmussen to John E. Landon, January 24, 1964). An insurance waiver signed by the artist was enclosed with the letter. Overall cracks related to this flaking were still visible in the work during selection of the study group paintings.

The stress of drying can crack brittle, weak paint films, as well as crack adjacent paints or cause "blind" intralayer and interlayer cleavage in brittle or weak paints that lie unseen beneath the surface of a paint layer. Hofmann's practice of reworking paintings over several months led to a similar problem, as lack of adhesion between wet and dry paint layers creates opportunities for cracking, lifting, and cleavage.[97] One inventory erratum among the artist's papers identifies six works completed by Hofmann between 1953 and 1958 as completely "repainted" (Hofmann papers, [ca. 1904]–78). Underbound ground and paint materials and areas of blind cleavage offer little internal strength to support heavy paint layers or stressed fabric and wooden support materials. Vulnerabilities of these complex constructions were discussed even while the artworks were being created, including at a 1962 presentation by Pomerantz, later published under the title *Is Your Contemporary Painting More Temporary Than You Think?*: "The piling up of cement-like loads of pigmented matter on flexible fabric supports is frequently encountered in avant-garde paintings.... Often, due to the obvious incompatibility of the extremely flexible support and the extravagantly thick, rigid paint film, which contributes to even further sagging of the support, large scale cleavage [interlayer cracking] and losses result" (1962, 19).

Uneven drying of paint materials can also contribute to canvas distortion. Uneven thickness of paint or uneven distribution of materials within a paint layer can lead to uneven drying that warps the canvas support and places additional stress on canvas and paint layers where distorted and planar surfaces meet. Expansion and contraction of a canvas in response to shifts in humidity or temperature can also exaggerate the stresses imposed on neighboring materials and expand existing cracking and cleavage.

The physical action of painting can also stress the structure of an artwork. "Depending on his concept," recalled Lillian Kiesler, who was Hofmann's school administrator from 1932 to 1947, "sometimes [Hofmann's] movements, as he painted, were delicate and lyrical. Other times, he was caught in a frenzy of painting, his movements were

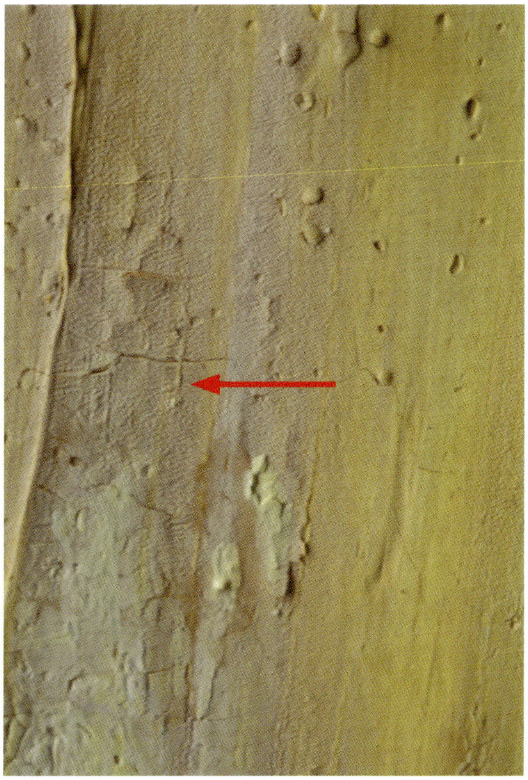
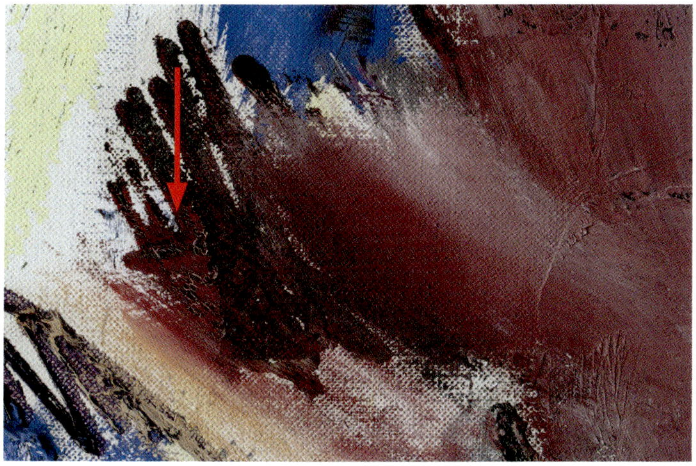
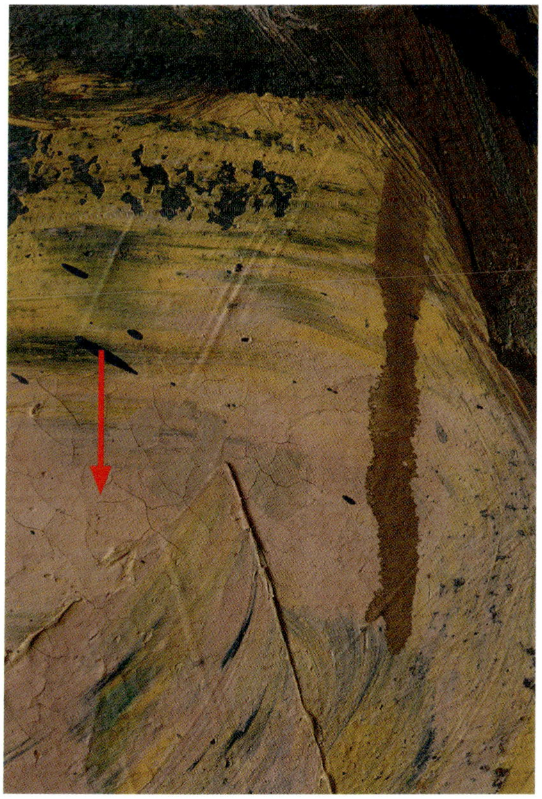
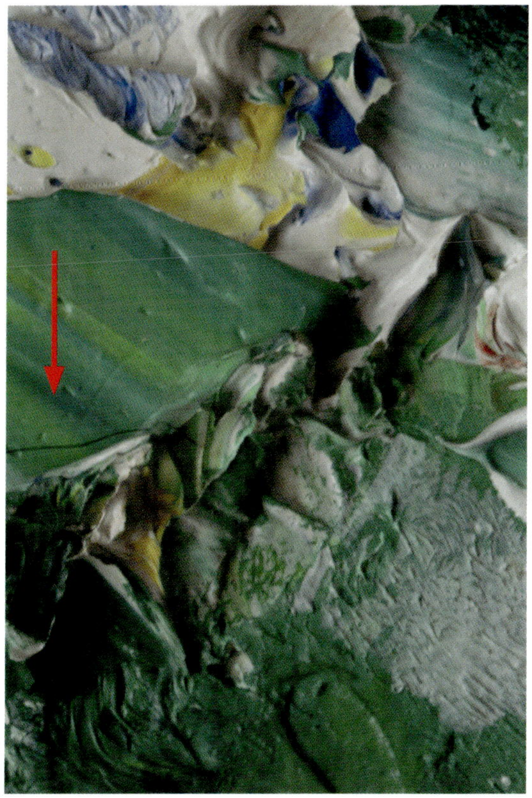

frightening....It gave me the aspect of a person spinning into space, but somehow the central axis was out of joint" (Kiesler papers, [ca. 1910]–2003, "As a Painter," 1). Elaine de Kooning reported Hofmann using his brush "like a dagger" (1950, 40), a practice that could distort the canvas support, crack dried paint materials, and expand existing cracking and cleavage. Such intensely physical painting practice left its mark on both compositional and preparatory layers, and the marks may appear immediately or over time. Environmental and mechanical stresses can also occur with normal handling or transport to and from storage and exhibition spaces. Similar to the drying cracks discussed previously, mechanically induced cracks in a paint film may remain stable, or they may create areas of weakness that transfer stress to adjacent materials, resulting in uneven stress throughout the painting. Newly applied paint or varnish materials can also seep through cracks in dried paint, altering the drying or aging characteristics of surrounding materials.

Paint Surfaces

The surfaces of Hofmann's late paintings offer distinct evidence of the artist's hand, as well as reveal condition issues that appear related to the artist's materials. Pinholes and burnished impasto, for example, often result from Hofmann's use (and subsequent removal) of paper compositional aids, and they are occasionally accompanied by small cracks in the upper paint layer that pose little threat to the structural integrity of the artwork (fig. 4.8). Similarly, evidence of the tape Hofmann used to edge his colored rectangles can be seen in altered paint surfaces and tilted impasto that remain following the tape's removal (fig. 4.9). Although Hofmann is said to have occasionally buried whole pieces of colored paper between paint layers (Rockwell 1982, 8), no adhesion gaps

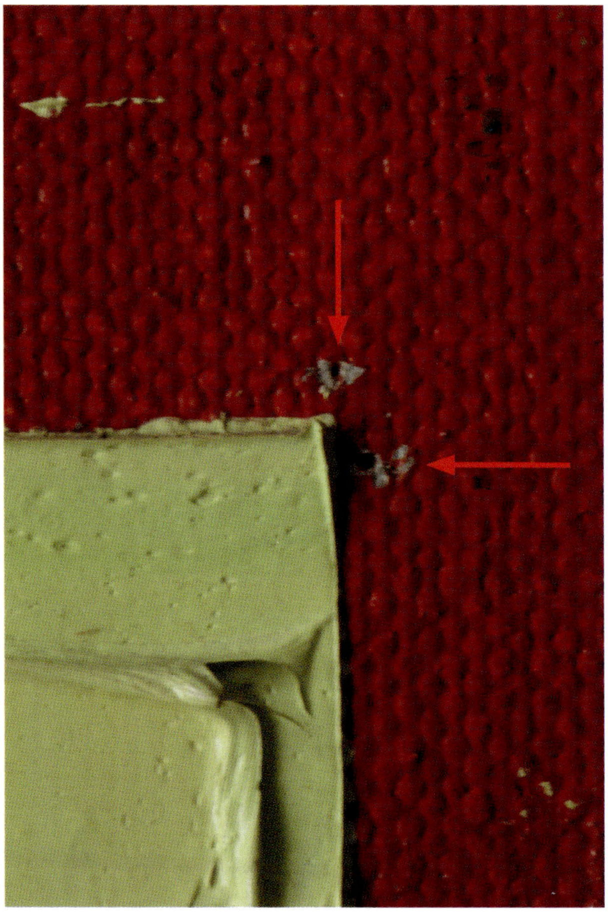

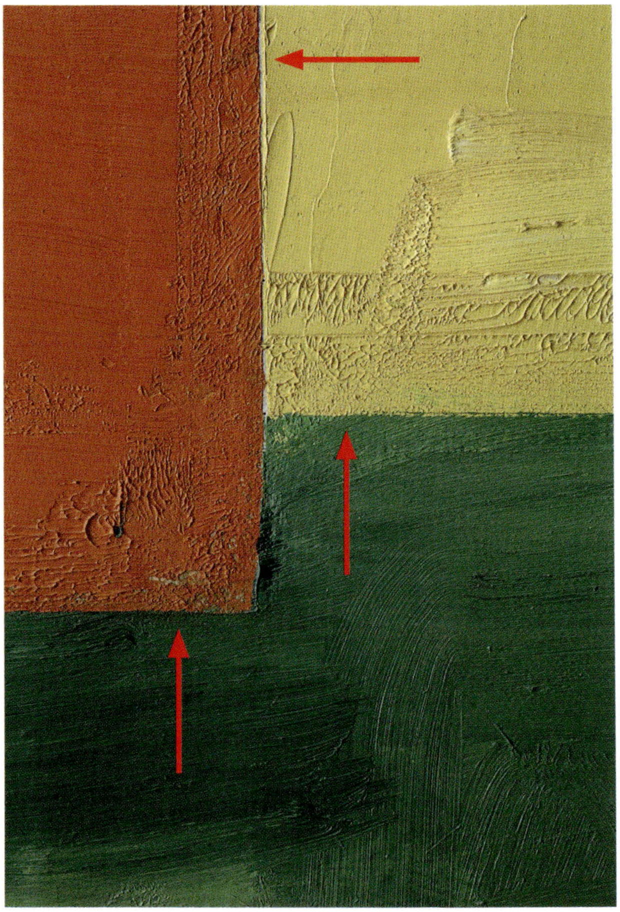

FIGURES 4.8 AND 4.9
Details of Hans Hofmann's *Magnum Opus* (1962; see fig. 6.9) and *Silent Night* (1964; see fig. 6.3), photographed in oblique light.

The images illustrate the limited cracking and paint loss associated with Hofmann's use of pushpins (left) and the altered paint surfaces resulting from tape removal along the edges of Hofmann's rectangles (right, borders of the orange and yellow rectangles).

FIGURES 4.10 AND 4.11
Details of Hans Hofmann's *The Vanquished* and *Indian Summer* (both 1959; see figs. 3.10 and 3.7), photographed with normal illumination. Surface wrinkling (top) and an uneven mix of materials applied directly from the tube (bottom) can retard the drying of thick paint layers.

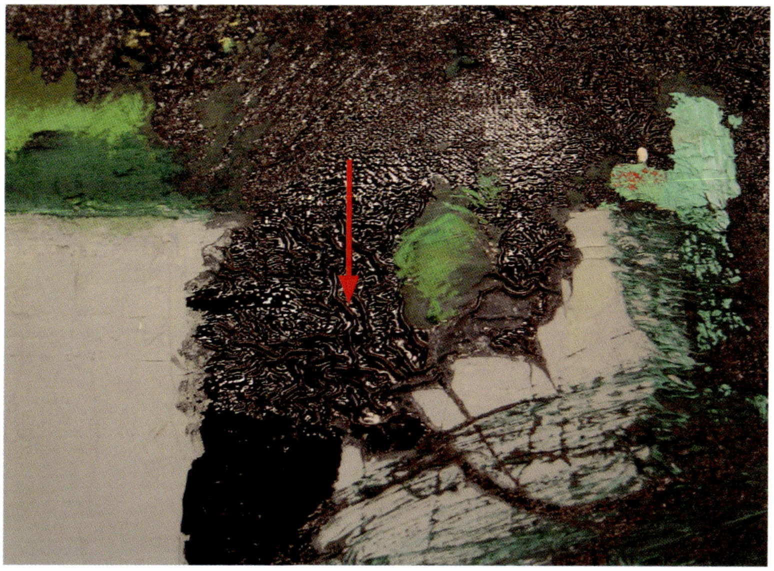

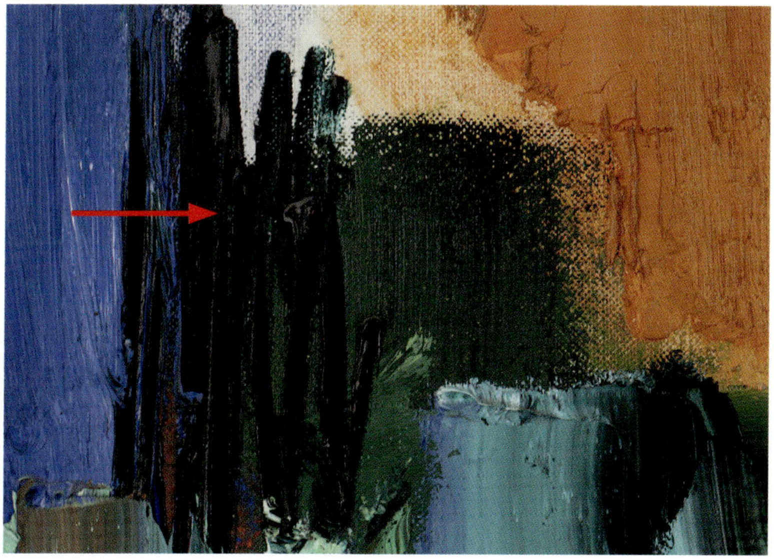

indicative of this practice are visible or have been recorded in the study group paintings.

Hofmann's works on canvas also exhibit surface phenomena related to the uneven drying of paint materials. A few examples of what appear to be poured black enamel-type paints are distinguished by surface wrinkling commonly associated with the expansion and contraction of fast-drying industrial paint formulations (fig. 4.10). Surface wrinkling is for the most part an aesthetic concern, although wrinkled paints can inhibit the continued drying of the underlying paint layer. Slowly drying, soft paints do not have the internal strength of fully dried paint films and can readily transfer stress to neighboring materials. Thick lines of paint applied directly from the tube (fig. 4.11) may also exhibit a delayed drying pattern, with the added complication of uneven composition. Pigment and binder can separate in the paint tube during storage, emerging as an uneven mix of oversaturated and overbound materials with uneven drying behaviors. In works for which Hofmann applied paint directly from the tube, excess binder from the paint could, but did not necessarily, spread to surrounding materials and create stains or additional regions of uneven paint composition.

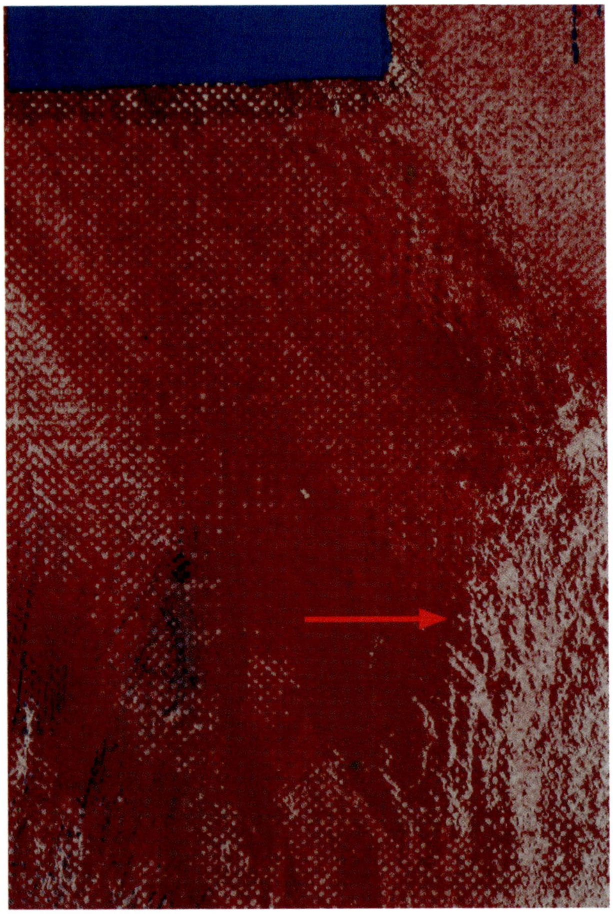

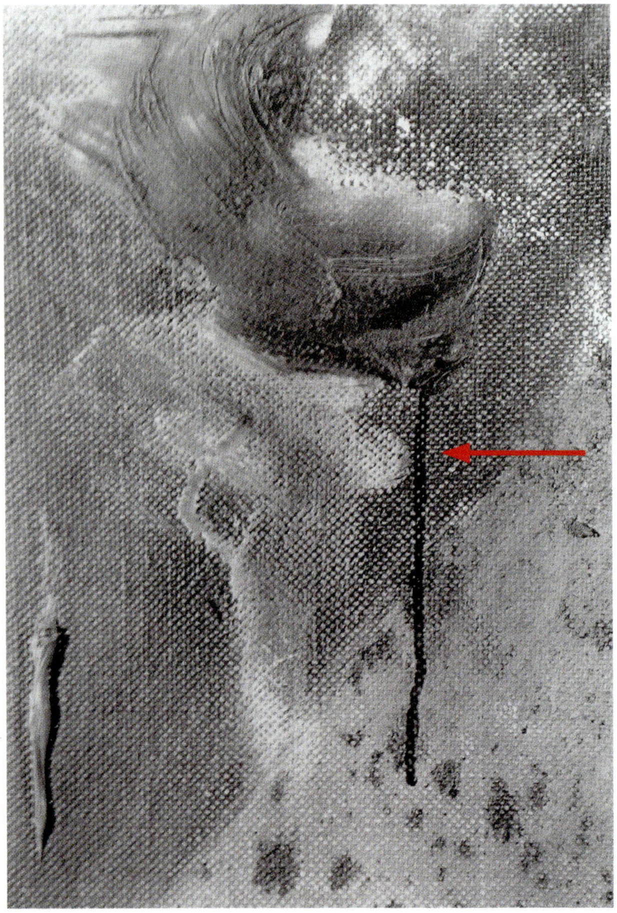

FIGURES 4.12 AND 4.13
Details of Hans Hofmann's *Polyhymnia* (1963; see fig. 6.10) and *The Clash* (1964; see fig. 3.11), photographed with normal illumination, showing surface phenomena related to the movement of paint materials in Hofmann's alizarin paints. Absorbent ground and paint layers can leach binder from thinned paints and cause cracking and desiccation (left), and migrating paint material can form liquid droplets on the surface of paint layers that drip or flow to other areas of the painting (right). The black-and-white detail, from Rockwell 1982, is labeled "Non-drying oil component running down from magenta paint. Sinking in of oil medium is evident in upper part of this paint area"; the original photograph has not been located.

The alizarin paints used in Hofmann's late work also exhibit underbound or oversaturated surfaces. "For the alizarin crimson applied to *The Third Hand,* 1947, with its absorbent gesso ground, and to *Struwel Peter,* 1965, with its porous alkyd ground," Rockwell noted, "the loss of oil medium from this lake pigment left it desiccated, cracked, crumbling, and flaking" (1982, 14). These underbound paint films (fig. 4.12) form poor bonds with surrounding materials, and exposed pigment particles on the painting's surface are susceptible to fading from exposure to ultraviolet radiation and atmospheric pollutants. According to conservation records, fourteen of the twenty-six study group paintings have exhibited underbound or desiccated paint; there are, however, no records of examination or analysis related to the potential fading of pigment color. Conversely, fourteen of the study group paintings have a history of soft or slow-drying alizarin paints: "Some paint surfaces remain tacky to the touch," noted Rockwell, with "sticky, almost fluid underpaint" (1982, 16). Four of the study group works continue to exhibit softened or sticky paint films.[98]

A condition issue characteristic of Hofmann's alizarin paints is exudation—the migration of mobile paint components within a paint film. Although the cause and mechanism of this migration is not clear, the migrating components appear on the surface of the paint film (and on film surfaces between paint layers) in forms ranging from crystalline fatty acids (Martin et al. 2013) to expressed liquid oils.[99] Crystalline exudates form a hazy film that obscures the underlying paint layer and hardens to a tough and opaque crust; liquid exudates form sticky masses that trap airborne contaminants and flow into or drip onto other paint materials. The exudation of liquid materials is commonly reported in association with Hofmann's late-career use of alizarin paints (fig. 4.13). "A few years ago, we received a distressed phone call from the [museum] registrar," Rockwell reported in the late 1980s, "informing us that a droplet of amber oil had been expressed from magenta paint on *The Clash*....The upper portion of the magenta area had grown chalky, while lower portions were progressively more saturated with oil" (1982, 16). Of the fourteen slow-drying paints noted above, eleven were alizarin-like colors that were soft and/or expressed liquid exudates.

Conservation History of the Study Group Paintings

Abstract Expressionist paintings can also be exposed to materials other than those chosen by the artist. These artworks also come into contact with material innovation through the act of preservation. In order to create an accurate catalogue of Hofmann's late-career materials and properly interpret the scientific analysis of samples from the study group paintings, it is necessary to differentiate Hofmann's materials from materials that may, for example, appear in samples as a result of conservation treatment. Treatment records and visual examination can be used to avoid most localized regions of conservation treatment during sampling, but samples taken from paintings that have undergone widespread or overall treatment are likely to contain nonoriginal materials that should be excluded from consideration. The conservation treatment records for the study group paintings assist this effort by providing information regarding nonoriginal conservation materials employed in the treatment of the paintings, but even the detailed conservation records can contain gaps in documentation or limited information regarding some conservation materials. It is important to take into account changing conservation methodology in the treatment of mid-twentieth-century paintings, including the introduction of removable synthetic materials, a growing emphasis on the artist's intent and the optical qualities of the modern paint surface, and a shift from overall to localized application of materials. While serving the preservation of Hofmann's material legacy, conservation treatment can impact the future appearance, analysis, and interpretation of these important works of art.

Inherent and chronic condition issues in Hofmann's late work—in conjunction with frequent requests for exhibition at other venues—have left many of the study group paintings with lengthy conservation assessment and treatment histories. According to the University of California's 1992 conservation grant application to the Getty Trust, "since their arrival on the UC Berkeley campus, the Hofmann paintings have required almost constant supervision" (SFMOMA records 1966–present). The earliest existing treatment records for the Hofmann Collection paintings are dated 1973, although there is evidence of earlier conservation intervention. New York–based paintings conservator Daniel Goldreyer examined several of the University of California's Hofmann Collection paintings in October 1966, and his treatment recommendations were documented by Lawrence Dinnean at the Berkeley Art Museum (then the University Art Museum)[100] in a handwritten note entitled "Goldreyer's Visit, 19 Oct 1966 / Examination w Mr, Mrs Goldreyer / MA Craft / L. Dinnean" (BAM/PFA records 1966–present). A penciled notation by Dinnean in the margin of a memo—written six years later by curator Brenda Richardson—noted that Goldreyer had treated several of the Hofmanns in the intervening period (BAM/PFA records 1966–present, "To: Larry / From: Brenda," May 25, 1972). It appears that Goldreyer was not the only person to work with the collection in its early years, as a November 1972 treatment proposal contains a penciled note by Dinnean that reads, "preliminary procedure already performed by Prof. Loran; Mr. Goldreyer is requested to examine, make conservation proposal, and perform further necessary conservation procedures."[101] Of the twenty-six paintings included in the study group, all have undergone at least one treatment campaign; seven paintings have undergone conservation treatment at least twice; and nine of the paintings have undergone conservation treatment three or more times.[102]

An overview of the primary conservation treatment techniques employed in the preservation of Hofmann's work is presented below, and a compilation of the conservation materials noted in the treatment records is provided in table 3 (see appendix 1). As analytical data cannot differentiate between similar adhesive or resinous materials utilized in different conservation procedures, conservation materials are grouped according to the procedures with which they are most associated.

Lining and Consolidation

Various adhesive materials may be found in Hofmann's paintings as a result of either lining or consolidation procedures. Lining is the practice of securing a painting onto a secondary support structure, usually performed to address canvas degradation or distortion that cannot be sufficiently reduced by other means. The secondary support may be a flexible material like canvas or a solid material such as a wood or aluminum panel. Early lining treatments typically adhered the lining material across the back of the entire painting; subsequent variations on the practice include edge lining and lining without adhesives.

Early lining treatments for Abstract Expressionist paintings did not differ significantly from lining treatments performed on paintings from other periods as recommended by conservators such as George Stout (1948), Morton C. Bradley (1950), and Caroline Keck (1954, 1962), and mentioned in earlier articles in the Fogg Art Museum's *Technical Studies in the Field of the Fine Arts* (published 1932–42; Stout was managing editor). Infusion of the painting canvas with wax-resin adhesive, coupled with lining to a solid wooden support was seen as a dependable alternative to earlier glue-paste lining systems for dealing with compound problems of lifting paint, blind cleavage, and distorted canvas. Such treatments were undertaken to address early problems with distortion and lifting paint in some of Hofmann's paintings. Infusion with wax-resin not only reestablished the bond between Hofmann's paint and support materials but also addressed extreme distortions in some paintings. "During the process of lining and panel mounting," noted Rockwell, "it is possible to eliminate some of the extensive planar distortions of the canvas which develop in many areas of [Hofmann's] thickly applied paint" and in areas where "the original canvas was stabilized by the wax-resin, which greatly reduced its dimensional responses to atmospheric humidity fluctuations" (1982, 13-A, 7). The adhesives used in lining are similar to the adhesives used to consolidate cracking or lifting paint films, a procedure that commonly precedes a lining treatment. Both traditional and modern adhesives may appear in localized or widespread regions of a painting, depending on the conservation procedure performed and the era in which the conservation treatment took place. Consequently, multiple adhesives may be present in samples from paintings on which multiple treatments were performed.

As lining techniques have changed, so too did the materials used in the lining of Hofmann's paintings. Many of the early wax-resin linings attached to Hofmann's paintings have been removed, and paintings lined to heavy Masonite panels have been relined to lightweight aluminum hollow-core panels or secondary canvas fabrics with synthetic adhesives that did not infuse the original canvas material or darken canvas or chalk-based paint extenders. Early synthetic-adhesive lining techniques still required the use of heat and pressure that could alter the texture and sheen of delicate paint surfaces and exposed canvas fibers or damage and distort thick paint layers and fragile impastos—a serious concern in the treatment of Hofmann's architectonic constructions. "Heavy, multilayered paint buildup ranges in thickness from ¼" to over 1"," recalled Rockwell, and "require[s] extensive impasto packing during its lining and panel mounting on the vacuum hot table" (1982, 9). Packing of the impasto involved localized use of paper fibers and aqueous adhesives: "One of the first concerns was the need to protect the delicate crests and points of paint impasto during treatment.... Protection was accomplished by packing moist paper fiber combined with aqueous adhesives around the vulnerable impasto with tweezers and fine spatulas.... After consolidation the pulp was softened again with moisture and carefully removed" (Rockwell 1982, 12).

Between the 1970s and 1990s, conservation methodology embraced low-heat or low-pressure lining techniques in response to rising concerns regarding the aesthetic and physical impact of invasive lining treatments on modernist paint surfaces (Villers 1974; Mehra 1981), and overall lining of Abstract Expressionist paintings became less prevalent. Goldreyer's 1986 treatment of a vandalized Barnett Newman painting in the collection of the Stedelijk Museum Amsterdam, for example, involved a full relining, but when the same vandal attacked another of the museum's works by Newman in 1997, museum conservators used synthetic adhesive and wire support threads to reweave threads in the painting and lining canvases damaged by the vandal, and they did not apply a new overall lining.[103]

Structural lining treatments are now infrequently employed in the treatment of paintings by Hofmann and his Abstract Expressionist colleagues. Examples of alternative lining treatments include the reweaving of torn threads, the deacidification of brittle canvas materials, strip lining (the practice of attaching strips of fabric to reinforce or replace the tacking edges needed to properly attach a painting to its stretcher) (Bustin and Caley 2003), "loose" lining (the stretching of a painting over a prestretched secondary canvas that relies on friction to create a bond between the primary and secondary fabric), and cami-lining—also known as stretcher bar lining—during which the painting remains on its stretcher and the secondary fabric is passed behind crossbars and attached to the inside edges of the stretcher. Cami-linings maintain limited contact between the primary and secondary support fabrics and rely on a pocket of air to cushion the original canvas during handling and transport (M. Mecklenburg 1991a, 1991b). By the mid-1990s, full linings no longer appear on treatment records for Hofmann Collection paintings; records of all works requested for loan show that, thereafter, they were required to be cami-lined before travel (SFMOMA records 1966–present).[104]

Varnishing and Inpainting

Similar groups of materials are utilized in conservation varnishing and inpainting procedures. As with lining and consolidation adhesives, the type of varnish applied to many Abstract Expressionist paintings has changed with evolving conservation research, and contemporary methodology guiding conservation varnishing treatments has become more cautious and more responsive to the desires of the artist.

Varnish may be used to resaturate and consolidate friable paint and to protect exposed pigment surfaces from environmental contaminants or light (when the varnish contains ultraviolet-blocking agents). Early varnishing treatments for Abstract Expressionist paintings did not vary significantly from the nineteenth-century practice of coating paintings with natural resin varnish before exhibition. As late as the 1970s, many institutional labs were still following the universal varnishing recommendations of Stout, Bradley, and Keck, who promoted varnishing as a preventive measure that "sealed" the paint layer against dirt and enriched the paint colors. "In many oil paintings of the twentieth century the effect desired by the artist is a mat or unshiny surface," Keck noted, but she asserted that "to leave these paintings without protection is to condemn them to an early death" (1954, 24).

In the 1950s synthetic varnishes rose in popularity as an alternative to the natural resin varnishes, which discolored over time and could be difficult to remove without affecting underlying paint layers containing similar materials. The low refractive index of some synthetic varnishes also produced a matte surface more in keeping with modernist aesthetics (Pomerantz 1962, 31). Early synthetic coatings were embraced by conservators—with mixed results. Low molecular weight synthetic resins were later found to exhibit soft surfaces susceptible to gathering dirt or shifting position at room temperature, a behavior also known as "cold flow."[105] Many popular varnishes were based on n-butyl methacrylate resins, which were found to form cross-linked brittle films that microfissured and appeared hazy, as well as exhibited solubility problems that limited subsequent removal of the film (Feller, Curran, and Bailie 1981). These varnishes, produced by different manufacturers, remained in widespread use until the late 1970s (Feller and Curran 1970; Lomax and Fisher 1990); more stable methacrylate varnish formulations emerged in the 1980s.

Synthetic varnishes were commonly applied to Abstract Expressionist compositions despite the artists' solidarity with their Cubist predecessors' rejection of varnishes. Unprimed canvases, popular with mid-twentieth-century artists, posed additional challenges when varnished: the varnish damaged delicate surfaces, captured and incorporated underbound pigment in the new surface coating, and saturated and darkened the exposed canvas support. Questions regarding the varnishing of content-laden modernist paint surfaces entered public discourse with a 1983 article by John Richardson entitled "Crimes against the Cubists," in which the critic proclaimed, "The surface of a [modern] painting *is* the subject—all the more reason to respect every detail of that surface."[106] The change brought on by varnishing unprimed canvases was so significant that color field painter Robert Motherwell famously bought back and destroyed one of his own paintings because it had been varnished. Similarly damaged, varnish-stained Kenneth Noland paintings at the Smithsonian Institution's Hirshhorn Museum and Sculpture Garden are the subject of a multiyear varnish removal research project. Throughout the 1990s, varnish removal campaigns for late nineteenth- and early twentieth-century modernist paintings were conducted at major institutions, including New York's Museum of Modern Art, which has thus far removed coatings from works in its collection by Picasso, Monet, and Matisse; the museum is currently removing varnishes from works by Abstract Expressionists such as Pollock.[107] Today, varnishes are commonly applied to Abstract Expressionist paintings as selective coatings, applied locally to replicate original surface characteristics or to resaturate degraded or damaged paint layers.

Changing attitudes toward the varnishing of modern paintings are reflected in the conservation history of the study group paintings. Varnish treatments common in the late 1960s were applied to Hofmann's work during this period, despite the artist's refusal to varnish his own paintings.[108] Dinnean's handwritten record of a visit to the Berkeley collection made by Goldreyer eight months after Hofmann's death includes the notation that "Goldreyer believes all Hofmanns should be varnished" (BAM/PFA records 1966–present, "Goldreyer's Visit, 19 Oct 1966"). The first archived correspondence regarding varnishing of the university's Hofmann Collection is an October 16, 1972, letter from Dinnean to Goldreyer requesting that the conservator "undertake the surface cleaning and sealing of a limited number of the 'White' Hofmanns during your visit to Berkeley this month" (BAM/PFA records 1966–present). Eight paintings were selected for varnishing to treat efflorescence and widespread flaking of paint, although evidence suggests that there had been previous varnishing campaigns. "Goldreyer has 'varnished' several of our Hofmanns," Dinnean penciled in the margin of the May 25, 1972, memo mentioned above, "over the past 6 or 7 years." The memo is a lengthy inquiry from curator Brenda Richardson to Dinnean—copied to museum staff including director Peter Selz and registrar Joan Feinberg—regarding the proposed varnishing project.[109] "I know that you believe varnishing is the only sure-fire way to preserve the Hofmanns," wrote

Richardson. "I also know that this is a controversial position.[110] The matter came up over a year ago in a curators' meeting, and there was a wide range of opinion, including Erle Loran's position to the contrary,[111] that Hofmanns should never be varnished" (BAM/PFA records 1966–present, "To: Larry / From: Brenda," May 25, 1972). Richardson's note reveals that Hofmann's opinion was known but that it was balanced against concerns arising early in the university's conservatorship of the artist's work. "We know of course that Hofmann himself did not want the paintings varnished," Richardson noted, "but I at least do not view that to be particularly relevant...if that were to imply their ultimate deterioration."

Richardson also suggested obtaining the advice of another conservator, "since institutions throughout the world may follow our methods, based on the fact that we have the largest single holding of Hofmanns anywhere and thus could be assumed to have expert judgment on this issue." Selz concurred in a memo the following day that the opinion of another conservator would be reassuring. "I agree with Brenda," he wrote, "that whatever we do will set a precedent for the treatment of Hofmann's work" (BAM/PFA records 1966–present, "To: Larry / From: Peter," May 26, 1972). No documentation exists in the curatorial archives regarding assessment of the paintings by other conservators, and an April 24, 1973, letter from Goldreyer to Dinnean notes that all eight paintings listed in the October 16, 1972, letter from Dinnean to Goldreyer were "sealed" with a poly(n-butyl methacrylate) varnish, later identified by Goldreyer as a mixture of DuPont's Lucite 44 and 45. "The proportions are varied," Dinnean noted. "They must always be mixed to suit the particular painting" (BAM/PFA records 1966–present, Dinnean to Goldreyer, October 16, 1972; "Nov. 72"; and undated note).[112]

By the mid-1980s, only localized varnishing was recommended for treatment of the Hofmann Collection. "None of the paintings should be varnished overall and none require linings at this time," San Francisco Museum of Modern Art conservator Jim Wright wrote to university curator Sidra Stich in 1986. "The variation[s] in the surfaces are important to the pieces....[Some] medium has sunken into the structure [and in] order to bring out the original harmony of the pieces, these areas could be [locally] varnished" (BAM/PFA records 1966–present, Wright to Stich, February 7, 1986). No varnish removal campaigns are noted in the treatment records for the Hofmann Collection paintings.

Resaturation and Exudate Removal

Some of the materials used in the procedures described above may appear in Hofmann's paintings as a result of restorative treatments intended to return the original balance of paint surfaces, or as a result of an effort to both saturate and consolidate friable paint. As discussed earlier, underbound paints—or those paints whose binding media have leached into other paint layers or have expressed onto the surface of the painting—develop rough surfaces that scatter light and cause colors to appear faded. A variety of replacement binding media are used by conservators for the resaturation of underbound paints. Locally applied varnishes surround underbound pigment particles, smoothing the surface and increasing the visible depth and intensity of the colors, but the effect can diminish as the oxidizing varnish layer shrinks and conforms to the paint layer's rough surface or as varnish materials are removed during subsequent conservation treatment. An alternative solution is to replace the missing oil or synthetic binding components of the paint mixture. This treatment resaturates more of the original material and is less prone to inappropriate surface sheen or solvent vulnerability than is local varnishing. According to Rockwell, the color of the underbound pigment has played a role in determining the proper saturation material for treating Hofmann's late-career works: "Stand oil, a particularly stable form of linseed oil...may be allowed to penetrate the medium deficient paint in several applications.... This works well for warm colors such as reds, oranges, yellows, and browns, where the slight yellow tone of the stand oil won't affect the color significantly. However, for cool colors and whites, polyvinyl acetate, a colorless, durable, synthetic resin, may be a better addition" (1982, 14).

A variety of materials may appear in modern paintings in relation to the conservation treatment of surface exudates, for which no successful treatment protocol has yet been established. "You will probably not be surprised," Dinnean reported in a letter to Goldreyer on March 12, 1973, four months after Hofmann's "White" paintings were varnished, "to know that a few areas of heavy impasto are continuing to bloom [exhibit hazy exudates]" (BAM/PFA records 1966–present). Local treatment with heat has shown some success in reintegrating crystalline materials into a paint film, but the procedure is not appropriate for all paintings, as heat can produce unexpected softening of paint or changes in the surface sheen of some paint materials.

Liquid exudates and crystalline solids can also be removed mechanically or with solvents, and while this treatment has less impact on the original paint surface or sheen and leaves little nonoriginal residue, there are concerns regarding the efficacy and long-term ramifications of this procedure—especially since the process must be repeated as materials continue to migrate to the paint surface. Recent conservation conferences have raised questions about the effect that the removal of exudates has on the overall composition of paint films, and researchers continue to investigate the behavior of exudates in oil paint films (Burnstock et al. 2007; Cooper et al. 2014). Varnish and wax barriers have shown some success in retarding the progress of exudates, although recent research has suggested that some exudates were only altered in appearance by treatment with varnish or wax and would therefore reappear over time (Martin et al. 2013). According to the treatment records, procedures utilized for recurring crystalline and liquid exudates in the study group paintings have included mechanical removal, removal with solvents such as benzine, varnishing with synthetic mixtures, and the local application of heat.

5 Identification of Hofmann's Late-Career Materials

The material constructions of Hofmann's last decade cannot be thoroughly understood without an examination of the materials themselves. If Hofmann's late-career paintings catalogue the breadth of his modernist influences, then we must understand how Hofmann's choice of materials reflected his early years in Europe and how the signature paintings he created in the United States advanced contemporary arts practice in an era art historian William Agee recently characterized as a "burst of investigation into the material world around us" (2012, 6). This chapter presents the first comprehensive identification and analysis of Hofmann's late-career materials. This large-scale compilation of materials data corroborates the anecdotal accounts and empirical observations reviewed in the previous chapter and sheds light on an aspect of mid-twentieth-century arts practice that directly impacts the condition of work produced by Hofmann and his Abstract Expressionist colleagues.

The goal of this materials analysis is to provide a significant and representative accounting of Hofmann's materials and then use that information to shed light on unique aspects of the artist's style and the aging behavior of his paintings. The large sample set precluded detailed identification of each material; this initial cataloguing of Hofmann's late-career palette is therefore based on the broad characterization of the binders and pigments in a large number of paint samples.

Analytical Methods and Sample Selection

Information regarding Hofmann's late-career materials was obtained using complementary analytical techniques that provided corroborating information. In every instance, the complementary information gained by the use of different techniques was consistent. Ground layer stratigraphy and identification of inorganic pigments were obtained with optical microscopy and scanning electron microscopy–energy dispersive spectroscopy (SEM-EDS). Fourier transform infrared spectroscopy (FTIR) and gas chromatography–mass spectrometry (GC-MS) were used to identify binders and organic pigments, and X-ray diffraction (XRD) was employed to distinguish between selected paints with similar inorganic components. GC-MS analysis was performed at the scientific laboratories of the National Gallery of Art, Washington, DC, and all other analysis took place at the scientific laboratories of the Museum Conservation Institute at the Smithsonian Institution.[113] An overview of the analytical techniques employed in this research appears in table 4, and table 5 tallies the number of separate analysis locations examined on each painting (for the tables, see appendix 1).

A total of 284 single and layered paint samples and 28 fiber samples were removed from the study group paintings and palettes. Compositional paints were sampled from each painting; support materials and preparatory layers were sampled where possible. When possible, samples were taken of every paint material appearing on every study group painting. Sampling of actual palettes from the artist's studio was limited to priming materials and examples of unusual colors or colors used to obtain color mixes. Sample sites were chosen to limit visible alteration of the composition and, when possible, to avoid noticeable areas of previous conservation treatment. All samples were stored in inert glass vials to avoid contamination from storage materials prior to analysis.[114]

Eighty-three of the paint material samples contained stratigraphy that made a total of 235 additional paints available for analysis. A total of 519 discrete paint layers from the study group paintings were analyzed for this research. Samples were obtained from the individual study group paintings on-site in the galleries, storage areas, or conservation facilities of their respective collections.

The implementation and results of each analytical method employed in this examination of Hofmann's materials are summarized below. The analytical methods are presented in the order of their application. Instrumental setup information is provided in appendix 2.

Optical Microscopy

Optical microscopy uses transmitted and reflected illumination to highlight different features in a sample material. Cross-section microscopy can provide information about stratigraphy and paint application (the blending of wet-on-wet paints, for example), as well as pinpoint the location (inter- or intralayer) of film failure. Ultraviolet-induced visible fluorescence of certain pigments and binders can be used to enhance stratigraphy, to provide basic identification (distinguish between protein and synthetic binders, for example, or elicit the sparkle of zinc under ultraviolet illumination), and to guide additional sampling and analysis.

Optical microscopy was utilized in this research primarily for initial examination and photodocumentation of the study group samples. All 284 single and layered paint samples and 28 fiber samples were examined and photographed under both normal and ultraviolet illumination. A total of 83 of the 284 paint samples presented multilayered stratigraphy and were selected for mounting and analysis as cross sections. The remaining paint samples were analyzed in their loose form.

Optical microscopy reveals a distinct late-career shift in Hofmann's preparatory materials. Reports of the artist's late-career preference for linen canvas are supported by examination of warp and weft fibers from the study group paintings. While the twisted, ribbon-like morphology consistent with cotton fibers is found in the warp and weft threads from both 1947 paintings from which samples were available (figs. 5.1 and 5.2), the segmented nodal structure associated with linen fibers is found in the warp and weft threads from all fourteen of the late-career paintings for which samples were available (figs. 5.3 and 5.4).

A shift in Hofmann's late-career ground and priming materials is also apparent. Under polarized light illumination, all samples of Hofmann's compositional white paints exhibit a speckled fluorescence common to zinc oxide pigment. While similar fluorescence is observed in Hofmann's early preparatory layers, it is not present in the ground and priming layer materials that appear in the artist's late-career works (figs. 5.5 and 5.6).

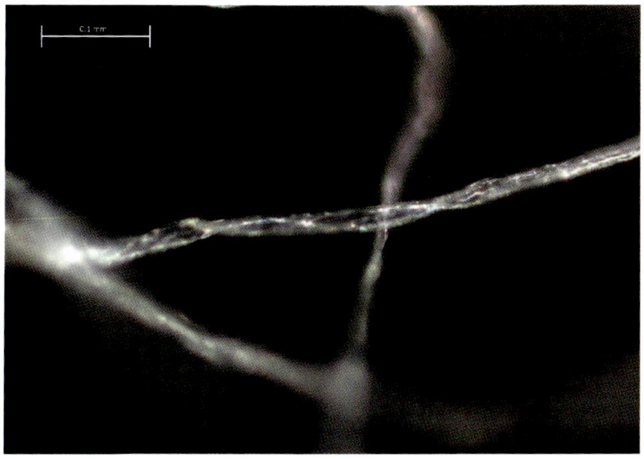

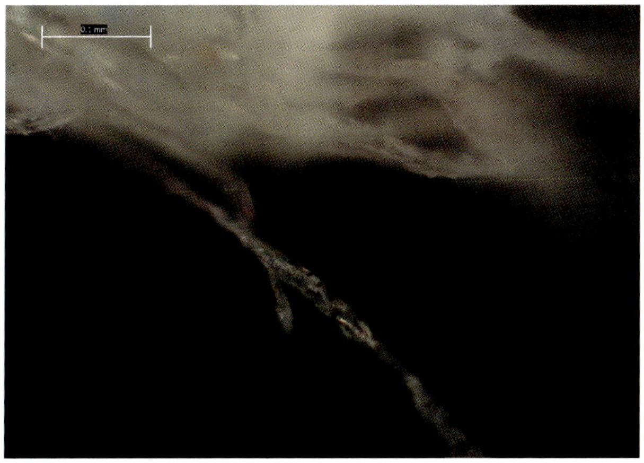

FIGURE 5.1
Photomicrograph of cotton warp fibers, 11.15× magnification (sample N17). Hans Hofmann, *Delight*, 1947. Oil on canvas, 127 × 101.6 cm (50 × 40 in.). New York, MoMA, 2.1956. Gift of Mr. and Mrs. Theodore S. Gary.

FIGURE 5.2
Photomicrograph of cotton warp and weft fibers, 11.15× magnification (sample Ecs04). Hans Hofmann, *Ecstasy*, 1947. Oil on canvas, 172.7 × 152.4 cm (68 × 60 in.). BAM/PFA, 1963.2. Gift of Hans Hofmann.

Cotton fibers display a characteristic twisted infrastructure.

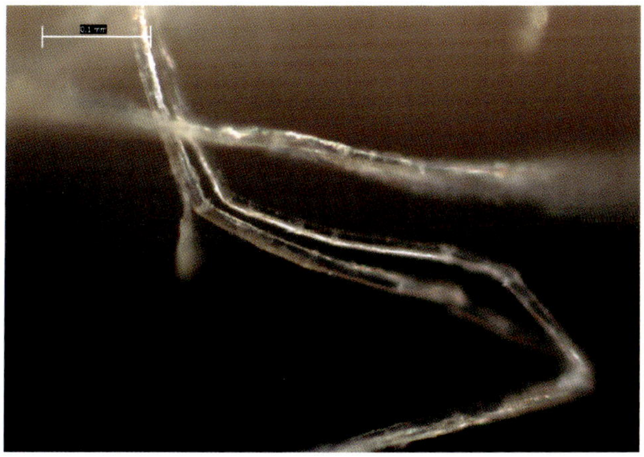

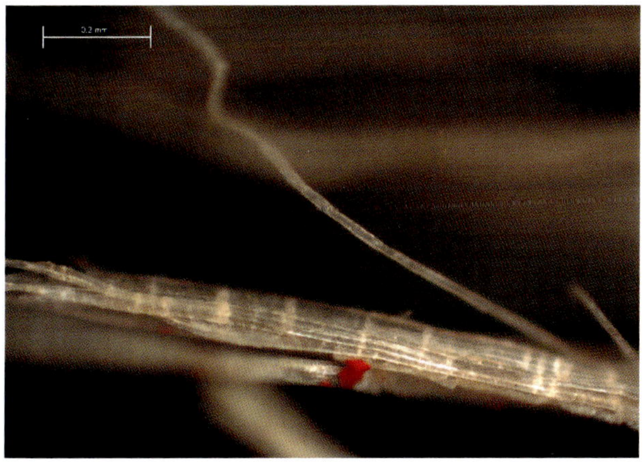

FIGURE 5.3
Photomicrograph of linen warp fibers, 11.15× magnification (sample R01). Hans Hofmann, *Ruby Gold*, 1959. Oil on canvas, 140.7 × 102.9 cm (55⅜ × 40½ in.). MAG, 60.37. Marion Stratton Gould Fund.

FIGURE 5.4
Photomicrograph of linen warp fibers, 11.15× magnification (sample N01). Hans Hofmann, *Memoria in Aeternum*, 1962. Oil on canvas, 213.3 × 183.2 cm (84 × 72⅛ in.). New York, MoMA, 399.1963. Gift of the artist.

Linen fibers display a characteristic segmented, or nodal, structure.

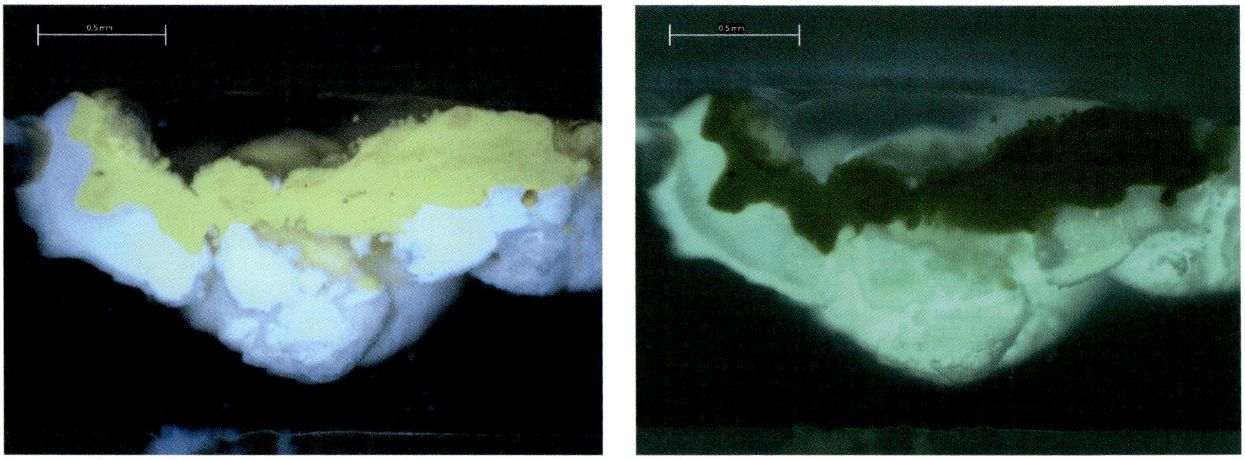

FIGURE 5.5
Photomicrograph of cross section containing compositional white paint, with dark field and ultraviolet illumination (355–425 nm), 2.79× magnification (sample C115). Hans Hofmann, *Above Deep Waters*, 1959. Oil on canvas, 214 × 132.1 cm (84¼ × 52 in.). BAM/PFA, 1965.13. Gift of Hans Hofmann.

The speckled fluorescence of zinc oxide pigment is visible in the compositional white layer.

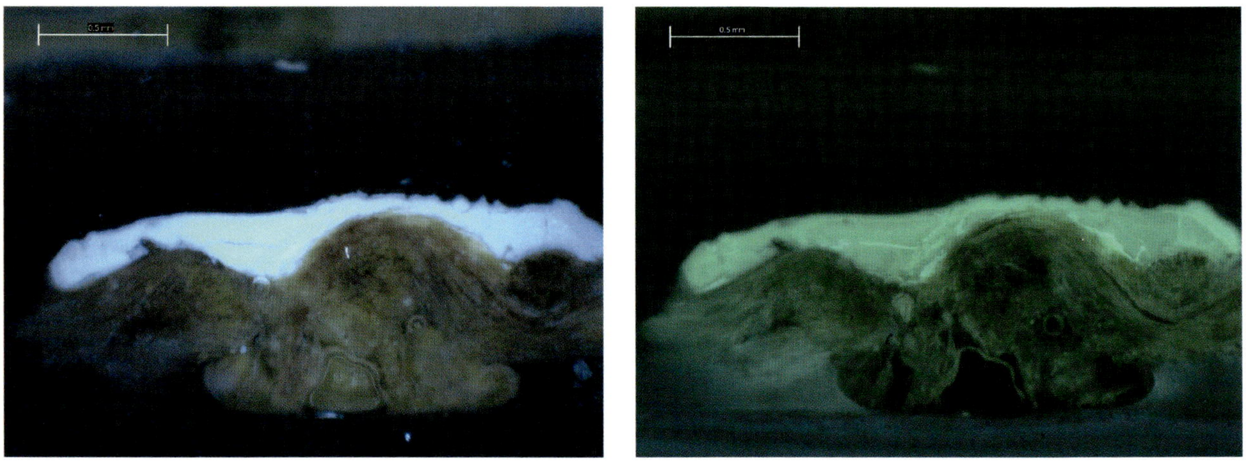

FIGURE 5.6
Photomicrograph of cross section containing commercial ground and artist priming layers, with dark field and ultraviolet illumination (355–425 nm), 2.79× magnification (sample C058). Hans Hofmann, *Tormented Bull*, 1961. Oil and enamel on canvas, 152.7 × 214 cm (60⅛ × 84¼ in.). BAM/PFA, 1963.6. Gift of Hans Hofmann.

No speckled fluorescence is visible in the preparatory layers of either fig. 5.5 or 5.6.

Pyrolysis–Gas Chromatography–Mass Spectrometry

Gas chromatography (GC) is a separation technique used to identify component materials in a mixture based primarily on differences in each component's volatility; pyrolysis (Py) is used to thermally fragment large component materials into smaller and more volatile components for analysis. GC is often coupled with mass spectrometry (MS), which uses high-energy electrons to ionize and fragment components and then sort the resulting charged fragments according to their mass-to-charge ratio and relative intensities. Material identification is then based on the resulting characteristic combination of ions. Py-GC-MS identification of modern paint materials should proceed with caution. The appearance of azelaic acid (a degradation product of oleic acid) in paint spectra is associated with aged paint films, but mid-twentieth-century commercial paint formulations include partially polymerized oils (the so-called boiled, or blown, oils), which give small azelaic acid peaks to the Py-GC-MS spectra of freshly applied paints. Similarly, derivatized phthalic acid (a GC indicator for alkyd paint) can also result from the transfer of commercial plasticizing agents used in many sample storage containers. The industrial paints favored by mid-twentieth-century painters typically contain a blend of different drying oils according to manufacturer preference, availability of raw materials, handling properties of individual pigments, or consumer application methods. Consequently, the palmitic:stearic acid MS ratios often used to identify specific drying oil materials cannot be viewed as definitive evidence of any particular drying oil within an industrial paint formulation.

GC-MS analysis was utilized in this research in order to differentiate between modern and traditional binding media in Hofmann's materials. A pyrolysis attachment was used for all GC-MS analyses to meet the thermal dissociation requirements of high-molecular-weight and heavily cross-linked polymers in modern paints reportedly used by Hofmann, including alkyds and modern industrial blends that contain alkyd or acrylic modifiers.[115] Samples of Hofmann's materials selected for Py-GC-MS analysis included those materials expected to contain modern polymer formulations—ground layer materials reported by others to contain house paints—and those paints exhibiting the flow characteristics or surface sheen associated with industrial paints. All white and black paints were analyzed with Py-GC-MS in order to differentiate suspected industrial paints in Hofmann's palette from traditional materials of similar color. GC-MS binding-media analysis was performed on 106 discrete paints.

Py-GC-MS analysis of Hofmann's paint materials confirms his preferential use of oil-based paint materials and his late-career shift to oil-based alkyd paints in his ground and priming layers. Py-GC-MS chromatograms of all 106 paint samples contain peaks consistent with the palmitic, stearic, oleic, and azelaic fatty acid components of oil paint; 27 chromatograms also contain peaks consistent with the phthalic anhydride component of oil-based alkyd paints (Bentley and Turner 1998, 61; Learner 2004, 57). This alkyd subset includes all of Hofmann's late-career preparatory layers and several of his splashed black paints. No alkyd paint indicators were identified in any of the colored paints analyzed with Py-GC-MS. No acrylic or water-based paint indicators such as polymeric acrylates or proteins were present in any of the samples analyzed with Py-GC-MS. While Hofmann's paintings do not present the conservation challenges of layered oil- and water-based paint media, degradation patterns in his work can be traced to the use of layered oil and alkyd paint materials (see chap. 6, "Discussion of Analyses").

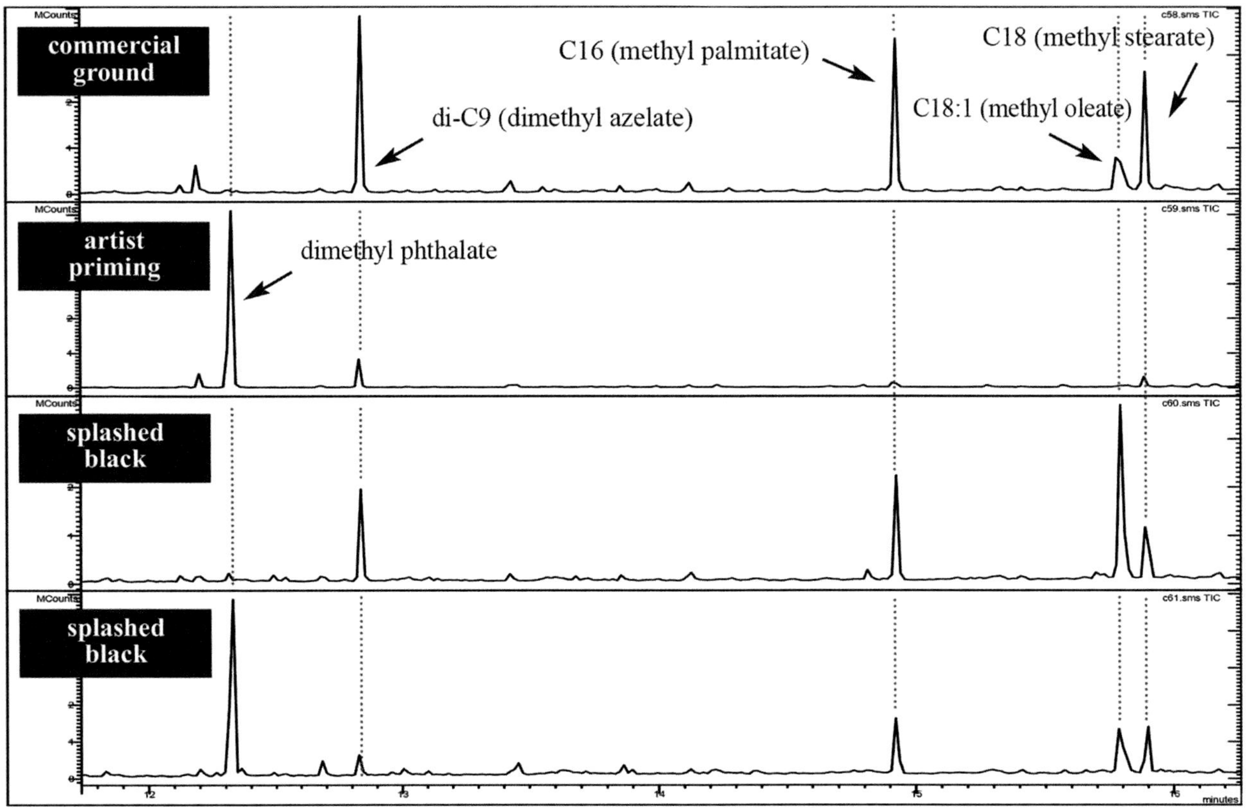

FIGURE 5.7
Gas chromatogram plots of methylated paint samples from Hans Hofmann's *Tormented Bull*, 1961. Dimethyl phthalate is indicative of the phthalic anhydride component of oil-based alkyd paints; dimethyl azelate, methyl palmitate, methyl oleate, and methyl stearate are the methylated fatty acid components of oil binders. Dimethyl phthalate is present in Hofmann's applied priming layer (sample C059) but not in the underlying, commercially applied ground layer (sample C058). The splashed black areas of the composition contain both oil paint (sample C060) and alkyd paint (sample C061). Similar combinations of oil and alkyd black paints are also visible on other study group paintings (see figs. 6.7 and 6.8).

White and Ground/Priming Layers

Py-GC-MS analysis of the study group samples confirms anecdotal reports of alkyd paints in the artist's late work. Py-GC-MS analysis of all three of the study group paintings produced by Hofmann in 1947 gave positive results for oil grounds, while chromatograms from all twenty of the late-career paintings and palettes from which ground or priming layer samples were available exhibited peaks consistent with the oil-based alkyd paints. This tally includes a priming layer applied by Hofmann over the commercially applied ground observed in the painting *Tormented Bull* (1961). Py-GC-MS analysis was not possible on the limited ground layer sample obtained from *Bald Eagle* (1960), but SEM-EDS analysis of that material (see below) reveals a bulked titanium white paint consistent with Hofmann's alkyd ground and priming paints and inconsistent with the artist's zinc white compositional paints.[116] Results were consistent among all methods.

Black and Colored Paints

Not all of Hofmann's splashed black paints are alkyds. Py-GC-MS analysis identified black paints with alkyd binding media on *Bald Eagle* (1960), *Tormented Bull* (1961), and *Heraldic Call* (1962), but other examples of splashed black paints on the same three paintings contained only oil paint binders (fig. 5.7). No alkyd binder was found in the splashed black paint on *The Vanquished* (1959). No alkyd binders were found in any samples of colored paints analyzed with Py-GC-MS. The large oleate peaks observed in figure 5.7 were commonly found in the chromatogram plots of paints containing zinc pigments; recent research by conservation scientist Christopher Maines (et al. 2011) linked high levels of oleate acids with the unique aging characteristics of zinc white oil paint layers in the work of Hofmann and his Abstract Expressionist colleagues (see chap. 6, "The Correlation between Condition Issues and Hofmann's Use of New Materials").

Other Materials

Py-GC-MS analysis provides some evidence of Hofmann's use of paint thinners or of subsequent conservation intervention. Although colored compositional paints were largely excluded from Py-GC-MS analysis of Hofmann's ground and priming layers, peaks consistent with the terpenoid components of paint solvents appear in Py-GC-MS chromatograms of a handful of colored paints from thirteen of the twenty-six study group paintings and five of the eight Trust palettes. Py-GC-MS chromatograms from eight of the study group paintings contained peaks consistent with the wax or acrylic polymer components of conservation consolidation or varnishing materials mentioned in the collection treatment records. Additional nonoriginal materials were later confirmed with FTIR analysis (see below).

Scanning Electron Microscopy–Energy Dispersive Spectroscopy

The resolution in a microscope is directly proportional to the wavelength of the light being employed; therefore, electron microscopes have much greater magnification and depth of field than optical microscopes. Images of inorganic materials present in a sample are produced from backscattered electrons by SEM in combination with the characteristic X-ray emission patterns of individual elements by EDS. This type of imaging technique can help answer questions about paint stratigraphy and artists' techniques or help differentiate between optically similar paints (white lead carbonate and white lead sulfate, for example). In combination, SEM and EDS can provide stratigraphic and elemental information in areas of paint degradation, or detailed information about pigment distribution and morphology. SEM-EDS can also differentiate application layers of the same paint (especially when surface dirt settles on one paint layer before the next layer is applied).

SEM-EDS analysis was utilized in this research to identify inorganic pigment families appearing in Hofmann's paints. EDS was utilized as a qualitative technique, as superficial pigment identification was sufficient to group the samples into inorganic pigment families and highlight potential organic pigments requiring additional analysis. Accurate selection of analysis location is aided by the increased resolution of the electron microscope and by elemental imaging that allows for more accurate discrimination between individual paint layers or individual paints within paint mixes. Backscatter electron images were taken of all 284 loose and mounted paint samples, and EDS inorganic materials analysis was performed on each of the 519 discernible paint layers within those samples. EDS analysis was performed on at least three disparate points within each paint layer.[117]

Table 6 (see appendix 1) presents a summary of the inorganic pigments identified in the study group paintings. Many of the pigment identifications obtained through EDS were confirmed through other analytical methods, such as FTIR (discussed below). Although sampling of the study group paintings was extensive, it is possible that additional pigments may be present. A limited number of paints were sampled on the Trust palettes; it is likely that colors in consistent late-career use also appear on the palettes.

White Pigments

The pigments identified in Hofmann's ground and compositional materials are consistent with the artist's shift from oil-based to alkyd-based ground layers. Mixed zinc and titanium white pigments consistent with zinc oxide oil paint formulations (distinguishing elements: Zn, Ti) were identified in the ground layers of all three of the study group paintings produced in 1947. No zinc pigments were identified in the ground layer materials of the seventeen late-career paintings from which ground layer samples were available, but all of the available late-career ground layer samples were found to contain a mix of titanium and calcium components consistent with the formulation of house paints referenced in Hofmann's papers.[118]

EDS analysis of Hofmann's compositional paints also confirms accounts of the artist's preference for zinc white paints. Zinc-titanium pigment mixes consistent with zinc oxide oil paint formulations were found on all of the paintings and palettes from which compositional white paint samples were available.[119] No titanium white paints were identified in any of the compositional layers of the study group paintings or palettes. No lead white pigments were identified anywhere on the study group paintings or palettes.

Yellow Pigments

Cadmium yellow (distinguishing elements: Cd, S) is the primary yellow pigment appearing in Hofmann's late-career work. Cadmium yellow paint appears in nearly every study group painting and palette, alone or mixed with zinc-titanium white or ultramarine blue.[120] Yellow ocher (distinguishing element: Fe) appears infrequently throughout the study group. Cadmium yellow and yellow ocher both appear in the 1947 painting *Delight* and in the later works *Sommernachtstraum* (1957), *Morning Mist* (1958), *Combinable Wall I and II* (1961), and *Silent Night* (1964), as well as on the unnumbered Trust palette on glass. In *Combinable Wall I and II*, a secondary shade of cadmium yellow appears in the form of the pigment extended with calcium sulfate. In *Imperium in Imperio* (1964), a secondary shade of cadmium yellow is extended with barium sulfate.[121] Zinc yellow (distinguishing elements: Zn, Cr) appears twice in the study group. Cadmium yellow and zinc yellow both appear in *Ecstasy* (1947) and *Equinox* (1958). EDS analysis of the yellow paint in the 1963 painting *Polyhymnia* identified a combination of cadmium and possible trace elements.

Orange Pigments

Cadmium orange (distinguishing elements: Cd, S, Se) is the only orange pigment found in the study group samples. Cadmium orange paint appears consistently in the study group paintings and palettes beginning in 1954—usually alone, but sometimes mixed with zinc-titanium white. A secondary shade of cadmium orange appears in the form of the pigment extended with calcium sulfate in the 1959 painting *Ruby Gold*. In *Combinable Wall I and II* (1961), a secondary shade of cadmium orange appears in the form of the pigment extended with barium sulfate.

Red Pigments

Cadmium red (distinguishing elements: Cd, S, Se) is another cadmium pigment that appears consistently throughout the study group paintings and palettes. Cadmium red paints appear regularly in compositions containing other cadmium colors. In the 1959 paintings *Above Deep Waters* and *Indian Summer*, a secondary shade of cadmium red appears in the form of the pigment extended with barium sulfate. Cadmium red extended with barium sulfate also appears on Trust palette M536-53. Additional reds in *Combinable Wall I and II* (1961) are achieved through a mix of cadmium red with zinc-titanium white or with the titanium white ground material. Both cadmium red and oxide red-brown paints (distinguishing element: Fe) appear in *Sommernachtstraum* (1957).

Green Pigments

Cadmium also appears in the majority of Hofmann's green paints. Many of Hofmann's greens are a visible mix of cadmium yellow and ultramarine blue, although a more homogeneous green mix containing components of cadmium green (elements: Cd, Zn, S) also appears in the study group paintings and palettes beginning in 1953. The cadmium-containing green often appears along with phthalocyanine green (distinguishing elements: Cl, Cu), and may therefore be the phthalocyanine-containing "[permanent] green light" that appears on pigment lists among the artist's papers (Hofmann papers, [ca. 1904]–78). The cadmium-containing green appears alone on *Exuberance* (1955), *Sommernachtstraum* (1957), *Morning Mist* (1958), *Memoria in Aeternum* (1962), and *Polyhymnia* (1963). Phthalocyanine green appears alone on the 1947 paintings *Ecstasy* and *The Third Hand* and on the later work *The Clash* (1964). A secondary cadmium-containing green appears in *Above Deep Waters* (1959) in the form of the pigment extended with barium sulfate.

A handful of other green pigments appears sporadically throughout Hofmann's late-career work. Viridian, also known as chrome green (distinguishing element: Cr) appears infrequently. Chrome-based green was the only green pigment used in *Delight* (1947), but it was used along with cadmium and phthalocyanine greens on *Silent Night* (1964). Iron oxide green (distinguishing element: Fe) also appears infrequently. Iron oxide green pigment appears along with cadmium green in *Ruby Gold* (1959) and with cadmium and phthalocyanine greens in the 1964 painting *And Out of the Caves the Night Threw a Handful of Pale Tumbling Pigeons into the Light*. A possible cobalt green (distinguishing elements: Co, Zn) appears along with cadmium green in *Memoria in Aeternum* and along with cadmium, phthalocyanine, and chrome greens in *Silent Night*.

Blue Pigments

Ultramarine blue (distinguishing elements: Na, Al, Si) is the primary blue pigment appearing in Hofmann's late-career work. Ultramarine blue pigment appears in every study group painting except the 1961 black-and-white composition *Tormented Bull*.[122] Ultramarine blue paint appears both as a stand-alone color and as part of a color mixture, frequently in combination with cadmium yellow. Another blue pigment found in Hofmann's late-career work is phthalocyanine blue (distinguishing element: Cu). Both phthalocyanine and ultramarine blue paints were identified on three works from the 1940s and 1950s: *Ecstasy* (1947), *Le Gilotin* (1953), and *Morning Mist* (1958). Phthalocyanine blue also appears on the unnumbered Trust palette. Cobalt blue (distinguishing elements: Co, Al) appears in four study group works from the 1960s: *Bald Eagle* (1960), *Magnum Opus* (1962), and the two 1964 paintings *Imperium in Imperio* and *Silent Night*. Cerulean blue (distinguishing elements: Co, Sn) appears in only one study group painting, the 1958 work *Equinox*. Likewise, Prussian blue (distinguishing element: Fe) appears in only one study group painting, the 1957 work *Sommernachtstraum*. Prussian blue also appears on one of the Trust palettes.

Violet, Magenta, and Purple Pigments

Both inorganic and organic purple pigments appear in Hofmann's work. Cobalt violet (distinguishing elements: Co, P) appears numerous times throughout the late-career study group paintings, beginning with *Le Gilotin* (1953) and continuing through the last study group painting, the 1965 work *Struwel Peter*. A heavily extended violet pigment in *Ruby Gold* (1959) presented inconclusive EDS spectra but similar SEM morphology to that of cobalt violet (figs. 5.8 and 5.9), and after excluding information related to potential conservation materials, the FTIR spectra for the violet pigment in *Ruby Gold* matched the FTIR spectra of EDS- and XRD-identified cobalt violet pigments in *Indian Summer* (1959), *In the Wake of the Hurricane* (1960), and *And Out of the Caves the Night Threw a Handful of Pale Tumbling Pigeons into the Light* (1964).

FIGURE 5.8
Backscatter electron image of sample R06, an unidentified violet pigment from Hans Hofmann, *Ruby Gold*, 1959. Oil on canvas, 140.7 × 102.9 cm (55⅜ × 40½ in.). MAG, 60.37. Marion Stratton Gould Fund.

FIGURE 5.9
Backscatter electron image of sample C129, a pigment with visual similarities to sample R06 and identified as cobalt violet. Sample from Hans Hofmann, *Indian Summer*, 1959. Oil on canvas, 152.7 × 183.4 cm (60⅛ × 72¼ in.). BAM/PFA, 1965.11. Gift of Hans Hofmann.

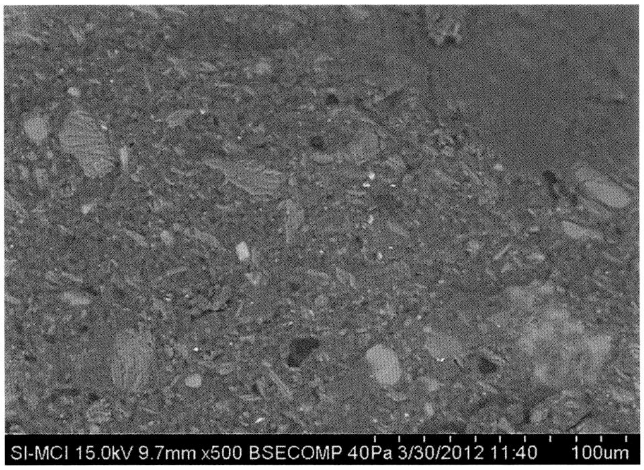

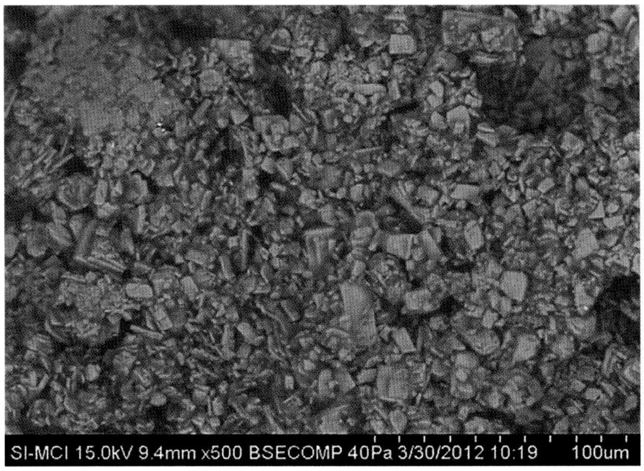

Ultramarine violet (distinguishing elements: Na, Al, S) appears only once in the study group, in the 1958 painting *Morning Mist*. EDS analysis of magenta paint samples from twelve of the twenty-three study group paintings identified levels of aluminum or barium consistent with mordant materials used in natural and synthetic dye-based colors. Further information regarding these pigments was obtained through organic analysis (see below).

Black Pigments

Hofmann's late-career paintings exhibit limited use of black pigments. Ivory black, also known as bone black (distinguishing elements: Ca, P), appears in eight of Hofmann's late-career study group paintings, beginning with the 1947 painting *Ecstasy* and including *Le Gilotin* (1953), *Above Deep Waters* (1959), *Bald Eagle* (1960), *Combinable Wall I and II* (1961), *Polyhymnia* (1963), and *The Clash* (1964). Mixtures of ivory black and phthalocyanine green were identified in all of these paintings except *Polyhymnia* and *The Clash.*[123] Ivory black pigment appears in both the oil- and alkyd-based black paints in *Heraldic Call* (1962). Neither the oil- nor the alkyd-based black in *Tormented Bull* (1961) has a discernable EDS signature, which suggests the presence of carbon black pigment, which is difficult to distinguish with elemental analysis. The alkyd paint in *Bald Eagle* may also contain carbon black pigment, while the oil-based black paint contains ivory black pigment. Potential carbon black pigments also appear on *Delight* (1947), the 1959 paintings *Ruby Gold* and *The Vanquished*, and *Bald Eagle* (1960). Carbon black pigment may be present in complex pigment mixtures or in black paints appearing alongside or mixed with any of the ivory black paints.

Fourier Transform Infrared Spectroscopy

FTIR analysis was utilized in order to identify organic pigments and other organic materials present in the study group paintings. FTIR uses radiation in the mid-infrared region (4000–600 cm^{-1} for micro-FTIR, 4000–400 cm^{-1} for FTIR in attenuated total reflectance [ATR]) to identify the chemical bonds within organic materials. Material identification is then based on the functional groups resulting from characteristic combinations of these bonds. The identification of modern paints using FTIR is complicated by the complex paint formulations of the industrial paints favored by mid-twentieth-century artists, which included experimental early paints and the use of other paint materials as modifying agents. Alkyd resins, nitrocellulose, and vinyl polymers, for example, are common plasticizers for acrylics, and oil-soluble additives can appear in non-oil formulations (Learner 2000).

A total of 248 loose and mounted samples from the study group paintings were analyzed with FTIR. Each of the existing 83 embedded cross-section samples was repolished to remove the carbon coating from previous SEM-EDS analysis and examined with an FTIR microscope attachment (micro-FTIR). FTIR information for small paint samples in 7 of the cross sections was overwhelmed by the spectra of organic components in the surrounding epoxy mount medium and thus provided no useful data. Useful FTIR data was acquired from 76 of the repolished cross sections. Loose paint samples were examined with the use of an ATR-FTIR platform, which incorporates infrared beam reflection from the surface in contact with the sample in order to allow examination of solid paint samples without additional preparation. Carbon coating could not be sufficiently removed from the loose samples previously used for SEM-EDS analysis; therefore, FTIR analysis of loose samples was performed on excess sample from 167 loose paints, as well as on excess sample of the same materials appearing in the five cross sections deemed too small for micro-FTIR.

FTIR analysis of samples from the study group paintings is limited to the identification of materials within the overall set of paints used on a single painting or palette. The microscope aperture required to produce adequate FTIR spectra from mounted cross-section samples with multiple paint layers cannot produce separate spectra for the individual paint layers. The information provided through FTIR, therefore, cannot be used to site organic materials within a particular paint layer. Similarly, loose paint samples are crushed during ATR-FTIR analysis, and if careful separation of individual paint layers is not performed in advance, composite spectra of different materials contained in the sample may be produced during analysis. Additionally, the refractive index of each paint material affects the penetration depth of the infrared beam, and therefore the materials themselves may cause variations in the physical depth from which information is gathered during analysis. This phenomenon impacts the analysis of both loose and cross-sectioned materials. While a large number of samples were analyzed for this research with FTIR, the limited number of analyses performed on each individual sample dictates that the FTIR analysis of Hofmann's materials be viewed as a qualitative, not a quantitative, technique.

Identification of organic pigments in Hofmann's work is based on the comparison of FTIR spectra from study group samples with the organic materials databases of the Infrared and Raman Users Group (IRUG), including IRUG standards for the epoxy medium used to embed cross-section samples and IRUG standards for conservation materials noted in the treatment records for the study group paintings. In all cases, FTIR results were consistent with those obtained from other analytical methods. Pigments identified in the study group paintings are organized below according to the categories presented in a contemporaneous edition of the international *Colour Index* (CI; 1957). Summaries of the organic materials identified in the study group paintings are provided in table 7 (pigments) and table 8 (conservation materials) (see appendix 1).

Organic Pigments—Blue and Green

Two phthalocyanine-based pigments were identified in the study group paintings. The *Colour Index* (CI) categorizes phthalocyanine colors as organic pigments, defined as "concentrated organic colouring matters containing no salt forming groups" (1957, 3267). Phthalocyanine pigments contain both organic and inorganic components and can therefore be identified using multiple analytical techniques—in the case of this study, by both FTIR and SEM-EDS (results noted above). FTIR analysis identified organic materials consistent with IRUG standards for the phthalocyanine-based CI Pigment Blue 15 (PB15, also known by the CI numerical designation 74160) in samples from two paintings—*Le Gilotin* (1953) and *Morning Mist* (1958)—and from two of the Trust palettes. Samples of pre-1953 phthalocyanine paints in Hofmann's work were either too heavily extended or contained too small a sample to be confirmed through FTIR analysis, although samples of phthalocyanine blue in the 1947 painting *Ecstasy* exhibited SEM-EDS peaks for copper consistent with those paints identified as phthalocyanine blue by FTIR. The designations PB15.1 through PB15.6, some of which are noted in table 7, reflect the related IRUG spectra for the pigment's standard crystalline variations (Herbst and Hunger 2004, 8, 418–45).

FTIR analysis identified organic materials consistent with IRUG standards for the phthalocyanine-based CI Pigment Green 7 (PG7; also CI 74260) in paint samples from ten paintings—*Le Gilotin* (1953), *Scintillating Space* (1954), *Sommernachtstraum* (1957), *Equinox* and *Morning Mist* (1958), *Above Deep Waters* and *Indian Summer* (1959), *Combinable Wall I and II* (1961), *Imperium in Imperio* and *Silent Night* (1964)—and from two of the Trust palettes. Samples from three additional paintings and one additional palette had SEM-EDS results that supported the identification of phthalocyanine green paints on a total of fourteen paintings and three palettes. The nonpolymorphous pigment appears as a single column in table 7.

Lake Pigments—Alizarin

Alizarin is one of two magenta colors that appear in Hofmann's late-career work. The *Colour Index* categorizes alizarin as a lake pigment, defined as "broadly similar to toners [see definition below] in type but precipitated in the presence of a substrate such as alumina, alumina blanc-fixe etc., the substrate being a necessary and integral part of the product" (1957, 3267). FTIR analysis identified materials consistent with IRUG standards for CI Pigment Red 83 (PR83, also CI 58000)—an anthraquinone-based alizarin substitute for natural madder pigment—in the 1947 painting *The Third Hand*, as well as in three later works: *Le Gilotin* (1953), *Morning Mist* (1958), and *The Clash* (1964). Samples of magenta paints from *Delight* (1947), *Ecstasy* (1947), *Magnum Opus* (1962), and *Polyhymnia* (1963), while too small to definitively identify the synthetic organic pigment used, provided FTIR matches to other samples identified as PR83 or exhibited SEM-EDS results consistent with an appropriate alumina substrate. No examples of PR83 were found on the palettes.

Toners

A second magenta color appears exclusively on Hofmann's late-career palettes. The *Colour Index* categorizes this rhodamine color as a toner, defined as "concentrated colouring matters produced by reaction of a water soluble dye with an appropriate precipitant" (1957, 3267). The xanthene-based rhodamine pigment known as CI Pigment Red 81 (PR81, also CI 45160) does not appear in any of the study group paintings, but FTIR analysis identified materials consistent with IRUG standards for PR81 in all three magenta paint samples available for testing from the Trust palettes. It is worth noting that alizarin (PR83) was not identified on any of the palettes from which samples were available.

Conservation Materials

FTIR analysis also revealed the presence of materials consistent with conservation records for the study group paintings. Table 8 provides a summary of conservation materials identified in the study group samples; the listing is not comprehensive, as sampling was designed to avoid nonoriginal materials. Materials consistent with IRUG standards for waxes appear in samples from four paintings with conservation records of overall or local wax infusion: *The Third Hand* (1947), *Scintillating Space* (1954), *Above Deep Waters* (1959), and *The Clash* (1964). Binding media consistent with IRUG standards for the poly(vinyl acetate) components of consolidation or inpainting materials[124] appear in only one paint sample, from an area of compositional white paint in the 1958 painting *Equinox*, noted in conservation records as having undergone local consolidation with poly(vinyl acetate). Acrylic media consistent with the IRUG standards for methacrylate varnishes appear in 43 samples from nine paintings with conservation records of local or overall varnishing treatment: *Le Gilotin* (1953), *Scintillating Space* (1954), *Equinox* (1958), *Bald Eagle* (1960), *Combinable Wall I and II* (1961), *Magnum Opus* (1962), and *The Clash, And Out of the Caves the Night Threw a Handful of Pale Tumbling Pigeons into the Light*, and *Silent Night* (all 1964).

The information provided by FTIR analysis regarding conservation materials is limited. The methacrylate polymers identified through FTIR analysis, for example, are a primary component of the poly(n-butyl methacrylate) varnish used by conservator Daniel Goldreyer in the early treatment of Hofmann's work, but they are also a primary component of the ethyl methacrylate methyl acrylate polymer coatings utilized in contemporary conservation practice. Nominal differences in the FTIR spectra of these materials (Learner 2004, 81) are difficult to discern when even small amounts of pigment or other material are present in a sample (fig. 5.10). The FTIR identification of

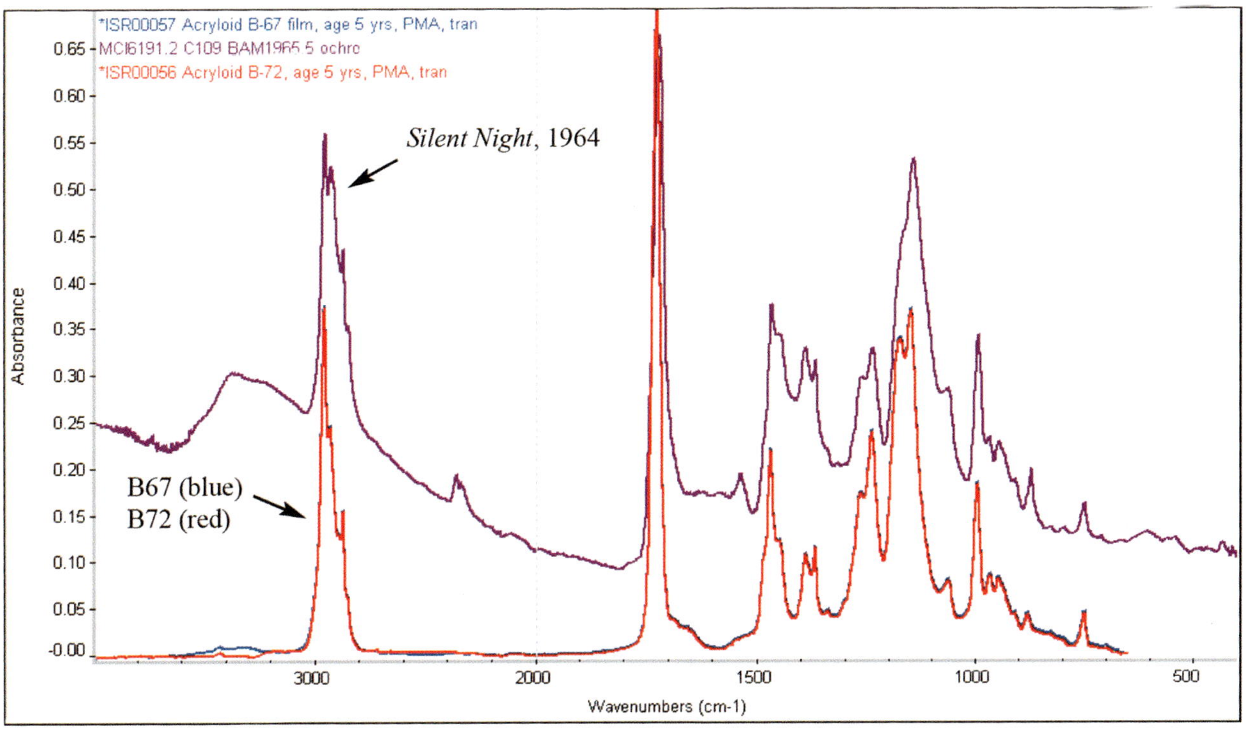

FIGURE 5.10
Comparison of Infrared and Raman Users Group (IRUG) spectroscopy standards for Paraloid B-67 (poly[isobutyl methacrylate]) and Paraloid B-72 (poly[ethyl methacrylate methyl acrylate]) with the spectra for sample C109, a varnish-containing ocher and cadmium yellow sample from Hofmann's *Silent Night*, 1964 (see fig. 6.3).

conservation materials in samples from the study group paintings is therefore limited to general categorization and cannot provide the level of identifying information necessary to date or differentiate conservation campaigns for the individual artworks.

X-Ray Diffraction

XRD analysis was utilized for a more detailed investigation of Hofmann's cadmium paints. In paint analysis, this technique can differentiate between pigments with the same elemental composition but different crystalline structures, such as anatase and rutile titanium. The characteristic diffraction of an x-ray beam by crystalline materials can be used to help date paintings executed in traditional artists' paints—rutile predates anatase, for example—but interpretation of such data must proceed with caution in the examination of modern industrial paints, which may continue to incorporate "outdated" artists' materials (rutile titanium creates a more stable exterior paint, but anatase titanium was included in some chalking, "self-cleaning" exterior paint formulations) (Rogala 2014, 80). XRD is particularly useful for differentiating between modern synthetic pigments of similar elemental composition, such as cadmium sulfide colors (ranging from yellow to red), chrome-based yellows, and copper-based greens, and it may help identify these pigments despite degradation-related color shifts. In modern paintings, XRD can also be used to analyze materials—such as oil paint fatty acids or acrylic paint surfactants—that have crystallized on the surface of a paint sample.

Unlike the analytical techniques discussed above—utilized for broad and complementary identification of a large number of samples—XRD analysis was performed in response to the discovery of calcium sulfate and barium sulfate extenders in secondary shades of yellow, orange, and red cadmium paints in Hofmann's paintings from 1958 onward (see "Scanning Electron Microscopy–Energy Dispersive Spectroscopy" above). Materials produce a characteristic diffraction pattern whether that material is present in a pure state or is part of a mixture, and this information about crystalline structure can be used to determine if different shades of a color were prepared utilizing the same pigmenting material. Identification of crystalline phases in the samples analyzed with XRD is based on a qualitative match provided by Jade 8.0 XRD software to reference patterns of known materials in the International Center for Diffraction Data (ICDD) libraries.

A limited number of pigment samples remained available for XRD analysis following all previous analyses. A total of 22 paint samples from the study group paintings were analyzed with XRD. The XRD sample group includes 11 examples of cadmium paint appearing on five paintings produced between 1958 and 1961—*Equinox* (1958), *Indian Summer*, *Above Deep Waters*, and *Ruby Gold* (all 1959), and *Combinable Wall I and II* (1961), along with 4 examples of cadmium paint appearing on three of the Trust palettes. The XRD sample group also includes 2 samples of paint whose identification through other analytical methods had produced limited or inconsistent results, 4 samples of pigment containing trace amounts of cobalt violet, and a single sample of a modern synthetic organic color. Because of the small number of samples analyzed, the limited number of XRD analyses performed, and the possibility that there were mixed paint materials within each sample, the interpretation of XRD data in anything other than a qualitative manner is precluded. A summary of XRD sampling and results appears in table 9 (see appendix 1).

Cadmium Sulfide Selenide and Cadmium Selenide Sulfide

A limited number of cadmium pigment polymorphs appear in XRD analysis of the study group samples. Out of the 15 cadmium paint samples, 13 exhibited characteristic diffraction patterns consistent with ICDD standards for the same cadmium sulfide selenide crystalline lattice.[125] Pigments identified as the same cadmium sulfide selenide polymorph include all 9 samples of red pigment and 4 of the 6 samples of orange pigment. For example, samples of both primary and secondary reds from the painting *Above Deep Waters* (1959) exhibit the same diffraction patterns, as well as match the diffraction patterns of the primary red and orange, secondary red, and pink pigments in *Combinable Wall I and II* (1961) (figs. 5.11 and 5.12). A different cadmium polymorph was found in only 2 of the pigment samples analyzed with XRD. Diffraction patterns consistent with ICDD standards for the same cadmium selenide sulfide crystalline lattice[126] appeared in conjunction with cadmium sulfate patterns in 2 orange pigment samples—one sample taken from the painting *Indian Summer* (1959) and one sample taken from a Trust palette. No other red or orange pigment polymorphs were identified in the study group samples.

The appearance of polymorph matches in Hofmann's paints provides interesting but ultimately inconclusive data. While XRD analysis reveals that the same crystalline cadmium pigments were mixed with different extenders to create primary and secondary shades of color in some of Hofmann's paintings (see EDS extender results above), there is no anecdotal or documentary evidence that Hofmann prepared any of his paints from raw pigments, and it is likely that the matching crystalline polymorphs identified in Hofmann's paints are an artifact of commercial paint manufacture. XRD analysis provides little insight regarding whether the extenders present in Hofmann's cadmium pigments were added by the paint manufacturers or by the pigment suppliers (Kremer 2013, 2014). An economical mixture of barium sulfates and cadmium sulfides —called "cadmopone" in contemporaneous sample books—was offered by pigment manufacturers during the 1950s and 1960s,[127] for example, but it is not in the scope of this study to trace the origin or function of the various extenders identified in Hofmann's cadmium colors. Paints with pigment polymorph matches cannot even be tied to a particular brand of artists' paint, because multiple manufacturers may have purchased raw or mixed pigment from the same industrial pigment supplier. Polymorph matches in paints obtained from paintings produced over several years (in this case, 1958 through 1964) may reflect an artist's extended use of personal paint stock, the extended distribution of commercial stock (through one store or many stores), or a lengthy run of production from a single raw pigment source, distributed to any number of secondary pigment or paint producers. Any one of these factors—or any combination thereof—may play a role in the limited number of polymorphs present in Hofmann's late-career cadmium colors.

Other Polymorph Identifications

Through XRD the identities of two previously unknown pigments were clarified. XRD analysis of an unknown blue pigment in the painting *In the Wake of the Hurricane* (1960) produced a diffractogram characteristic of ICDD standards for the lapis-based pigment lazurite,[128] and analysis of an unknown yellow pigment in the painting *Polyhymnia* (1963) produced a diffractogram characteristic of ICDD standards for cadmium manganese sulfide.[129] Trace levels of manganese in certain crystalline formations of cadmium sulfide yellow pigment may have produced the unique resonance responsible for the latter diffractogram (Dorain 1958), but the appearance of lazurite in Hofmann's work seems unrelated to any variation of previously identified pigments and is unique to this single study group painting.[130]

XRD analysis was also performed on cobalt violet pigment samples that proved difficult to identify through other analytical techniques. As with previous techniques, not

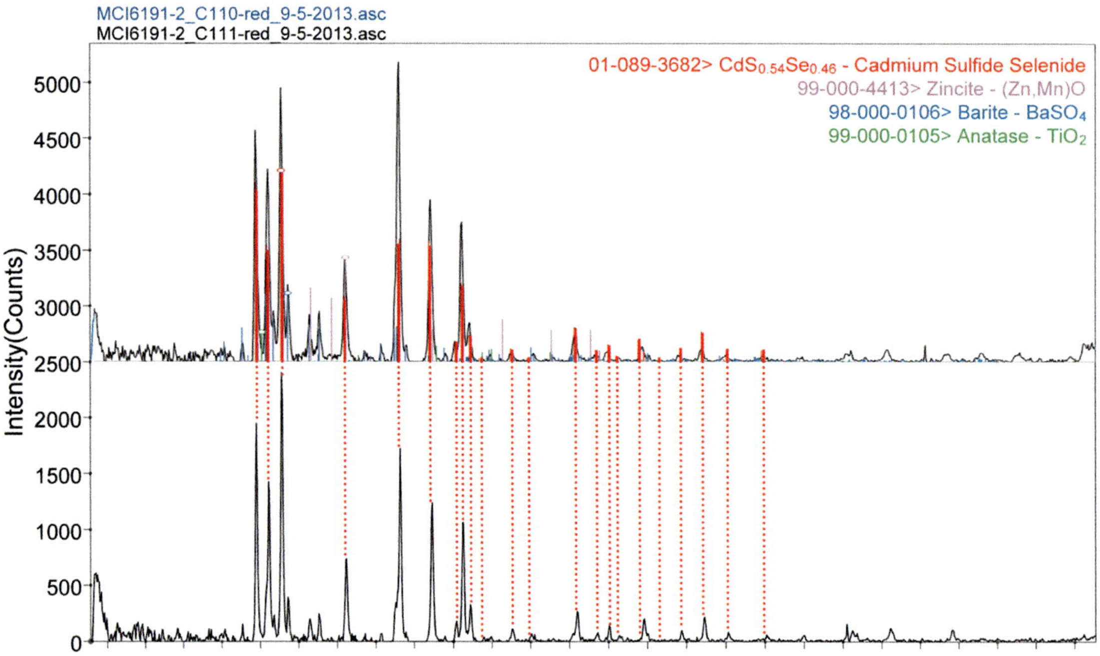

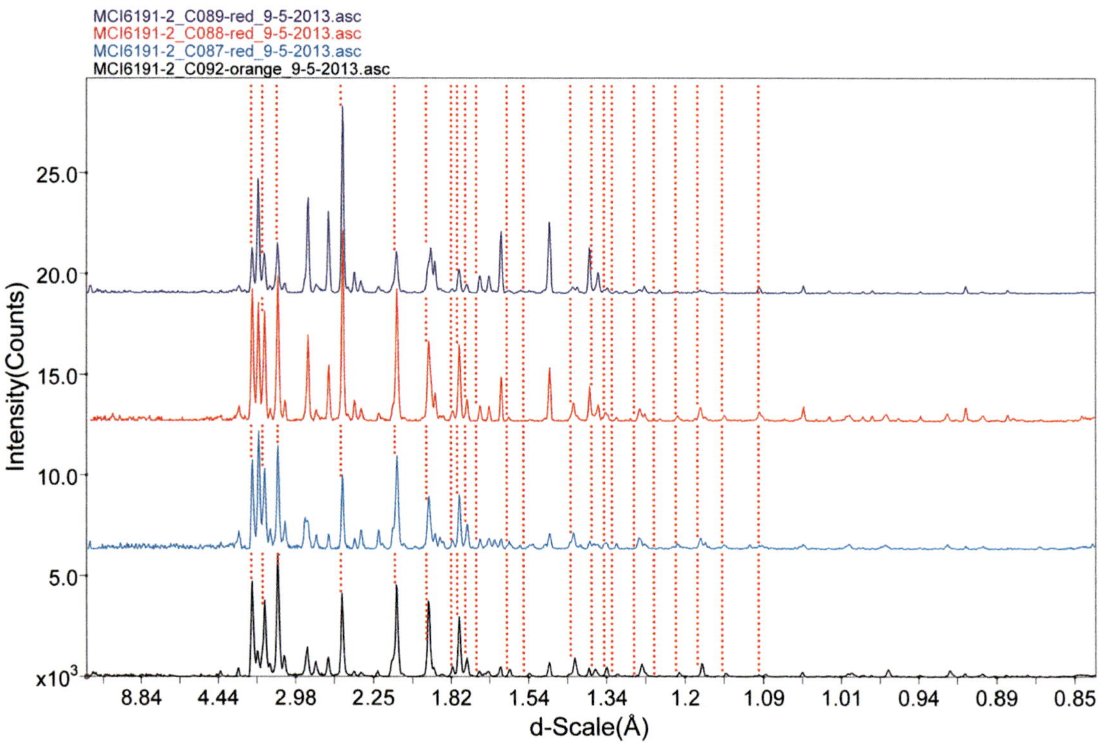

FIGURE 5.11
Diffractograms for primary and secondary red pigments (samples C110 and C111, respectively) appearing in Hans Hofmann's *Above Deep Waters*, 1959.

FIGURE 5.12
Diffractograms for primary red and orange (samples C087 and C092, respectively), secondary red (sample C088), and pink (sample C089) in Hans Hofmann's *Combinable Wall I and II*, 1961 (see fig. 3.4).

All 6 paint samples exhibit diffraction patterns characteristic of the same cadmium sulfide selenide crystalline lattice. The varying peak heights within crystalline phase matches may have various causes, such as crystal lattice orientation or damaged crystals, or they may be influenced by other crystalline phases within the pigment mix. The unmatched peaks represent diffraction patterns characteristic of the zinc white, titanium white, and extender components of the paint mixtures.

all samples of pigment previously identified through SEM or FTIR as cobalt violet produced diffractograms identified as cobalt phosphate, although all 4 samples analyzed with XRD did produce diffractograms consistent with ICDD standards for dittmarite, a crystalline composition of trace elements found in XRD analysis of cobalt violet by scientists at the Canadian Conservation Institute (Corbeil, Charland, and Moffatt 2002).

XRD analysis was also performed to confirm the absence of crystalline coloring materials in samples of organic pigments. XRD analysis of a pigment sample identified by FTIR as *Colour Index* Pigment Red 81 produced diffractograms solely for the rhodamine-based pigment's inorganic mordant materials.

6 Discussion of Analyses

> Sometimes the way out of [an] impasse in historical work comes from proposing another set of possible descriptions that the paintings in question might be seen to "come under."... What sorts of new orders in the objects would be set up, if we chose to look at them *this* way?
> —T. J. CLARK, "In Defense of Abstract Expressionism," 1994

Hofmann's most prolific period of creation is distinguished by a limited palette. A time line compiled from the analytical data reviewed thus far (fig. 6.1) reveals that Hofmann's most recognizable modern paintings were constructed from a small number of colors with only minimal use of mid-twentieth-century paint media. Apart from the artist's adoption of alkyd ground layers, the distinctive visual vocabulary of Hofmann's later years is fashioned largely from traditional artists' oil paints. The analytical data presented confirms anecdotal accounts of linen canvas and alkyd ground or priming layers in Hofmann's work, as well as the artist's preference for a bold but limited selection of colors. The modernist nature of Hofmann's color selection is explored below, along with the important relationship revealed between late additions to Hofmann's palette and recurring condition problems found in the artist's work.

Hofmann's Late-Career Materials

Preparatory Materials

New paint media appear primarily in Hofmann's preparatory layers. Shifts in Hofmann's materials from the late 1940s to the early 1950s recalled by students and confirmed by analysis include a switch from cotton to linen canvas supports and the introduction of alkyd paints in Hofmann's ground and priming layers. Alkyd paints also appear in Hofmann's sporadic use of splashed black alkyds, but unlike the artist's earlier constructions—which Thornton "Tony" Rockwell described as "several different types of media... combined in one painting including oil, casein, duco enamel, gouache, and India ink" (1982, 4)—Hofmann's selective compositional applications of alkyd paint are the artist's only late-career deviations from an otherwise conventional oil-based palette.[131]

Oil-Based Color

A handful of bright, bold colors is central to Hofmann's palette. Several principal colors in Hofmann's palette appear in the earliest study group paintings and continue in regular use throughout the rest of Hofmann's career. Zinc white is the only compositional white that appears on the later study group paintings, excluding trace mixtures and one instance

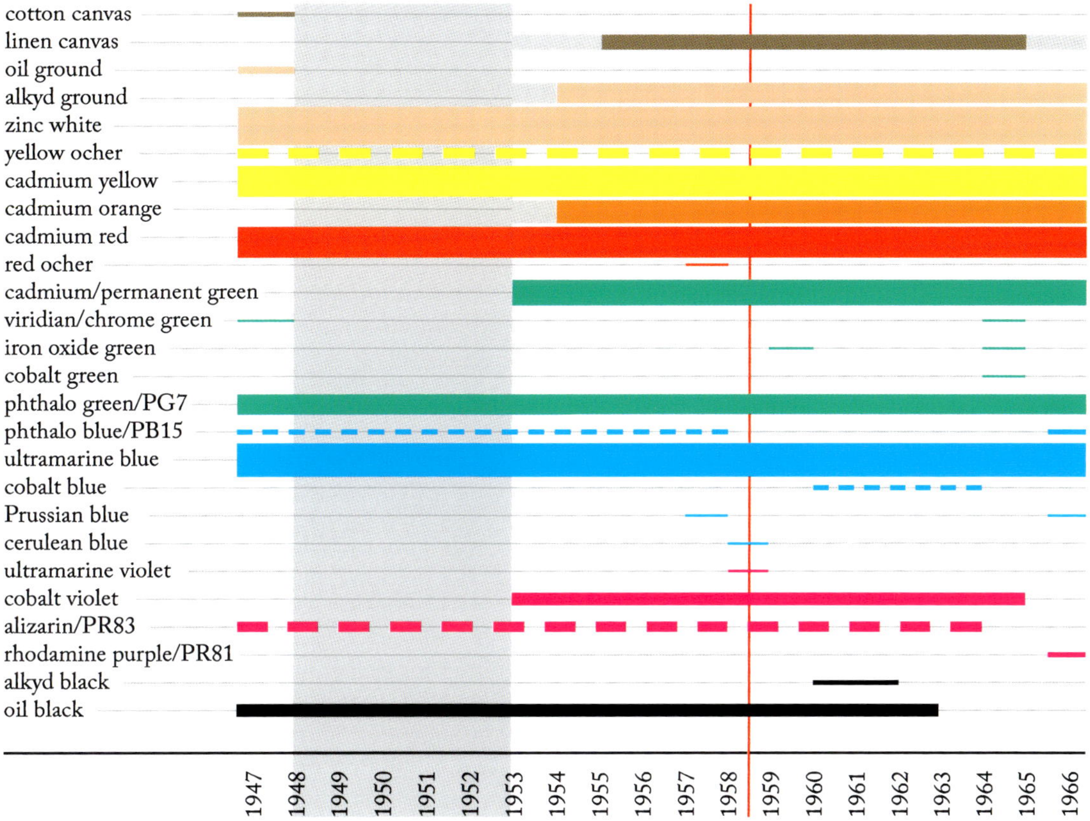

FIGURE 6.1
Time line spread of Hofmann's late-career materials. If use of a particular pigment is consistent, a solid line is drawn from beginning to end dates; sporadic use is represented by a dashed line between beginning and end dates. The thickness of the line denotes the relative number of paintings on which the paint is found within the study group. Not all colors were sampled from the study group paintings; colors not found in the study group works may exist in other paintings from the same year. Not all materials were sampled from each year in the time line; the starting and ending dates of common materials are extended with a horizontal gray bar to reflect their likely use. The vertical red line indicates the year Hofmann closed his schools and began to paint full-time. No study paintings appear within the vertical gray time frame.

of ground layer material used as a compositional paint reported by Tallent (1988). This is in keeping with accounts of Hofmann's disdain for lead white, a hazardous material still employed by colleagues such as Franz Kline (Rogala et al. 2010). Cadmium red is the only red pigment identified in the paintings, and although not employed in the same exclusive manner, cadmium yellow appears in every painting that includes yellow paint.[132] Ultramarine blue is similarly found in every blue-containing composition, either alone or among a selection of blue pigments. Alizarin crimson and phthalocyanine-based blue and green are among a handful of synthetic organic colors seen regularly throughout Hofmann's late-career work. These principal colors form the bulk of the materials identified through analysis of the study group samples.

Other compositional colors in common use by Hofmann in his later years appear in the study group paintings beginning in the early 1950s. Cadmium-based orange and green pigments appear during this period, and although archival lists and the anecdotal recollections of Hofmann's students allude to its earlier use, cobalt violet does not appear in the study group paintings until 1953. Other colors appear sporadically in the last decade of Hofmann's career—of note is the disappearance of alizarin crimson from the study group paintings after 1963 and the appearance shortly thereafter of a rhodamine-based magenta, although it is not clear whether this switch is in fact a substitution of one pigment for another and, if so, whether the switch was intentionally made by Hofmann or was a formulation change by a favored paint manufacturer. There is also a period in the early 1960s when cobalt blue temporarily replaced phthalocyanine blue in Hofmann's palette, but there is otherwise no noticeable shift in the artist's materials after 1958, the year Hofmann closed his schools and returned to full-time painting.

Hofmann's Palette and Modern Art History

The bright colors and new pigment formulations that dominate Hofmann's late-career palette are a reflection of the artist's years among the avant-garde communities of Europe and the United States. A chronology of twentieth-century modernism—and the role played by Hofmann's years in Munich, Paris, and New York—is visible in the artist's distillation of modernist color. The connections between Hofmann's colors and their associated modernist periods reveal the extent to which Hofmann's seemingly conventional palette was determined by modern art history.

The Colors of Early European Modernism

Modernist colors at the root of Hofmann's palette echo his time in Munich and Paris prior to World War I. Half of Hofmann's late-career colors are among the pigment formulations identified—by science writer Philip Ball and conservators from London's Tate and National Gallery—in the work of the Impressionist painters Hofmann studied as a young artist.[133] The lack of surviving works from Hofmann's early years as a painter precludes a definitive *terminus post quem* for these pigments, although Impressionist colors make up half of the handwritten pigment list that accompanied Hofmann's early years in the United States. The "push and pull" of color in Hofmann's signature work reflects not only the nineteenth-century color theories of Impressionist favorites Michel-Eugène Chevreul and Ogden Nicholas Rood[134] but also relies on the Impressionist palette of synthetic color formulations that Ball called "biased heavily towards…new materials" (2001a, 181). Colors from Hofmann's bold palette that were considered "modern" during his early years as an artist include zinc white, cadmium yellow, cobalt blue, viridian, and cobalt violet—a color featured in Rood's innovative treatise on complementary colors. "I have finally discovered the true color of the atmosphere," Édouard Manet is said to have proclaimed in 1880, the year that Rood's *Modern Chromatics, with Applications to Art and Industry* was published in Paris. "It's violet.…Three years from now everyone will work in violet" (Claretie 1881, 226; Ball 2001a, 184). Other principal pigments in Hofmann's palette that entered commercial production during the artist's lifetime include the striking cadmium oranges and reds embraced by modern painters during Hofmann's years in Paris. The "wild" color of experimental artists such as Henri Matisse and André Derain was dominated by newly accessible color formulations. "Once I had discovered that red color," said Matisse, "I put [other] studies in a corner, and that's where they'll stay.… I find that all these things—flowers, furniture, the commode—become what they are for me only when I view them together with that red" (Matisse 1995a, 62). The colors of Hofmann's Parisian colleagues became central to his later work. "As it now looks to me," professed Clement Greenberg shortly before Hofmann returned to full-time painting, "you could learn more about Matisse's color from Hofmann than from Matisse himself" (1957, 84).

The Materials of Mid-Twentieth-Century American Modernism

Hofmann's use of modern paint media has origins in both Europe and the United States, for although alkyd paints appeared in the compositions of Parisian colleague Pablo Picasso as Hofmann donned the mantle of modernist art teacher in Germany,[135] the use of alkyd priming layers took popular hold only after the artist was established in New York City. Harriet Standeven noted that this particular use of house paint among Hofmann's colleagues was often pragmatic: "Peculiar to the use of commercial paints during this earlier period…is that their presence was *implicit*, in that artists did not explore their unique physical capabilities, nor was their inclusion intended to make social commentary. Rather, commercial paints tended to be used for practical reasons: confined to priming layers or applied as though they were traditional artists' oils" (Standeven 2003, 190).

Similarly, the Permalba paint that appears throughout Hofmann's late-career paintings (identified through a proportional mix of zinc and titanium white pigments) likely came to the attention of Hofmann through its ubiquitous use by early American modernists including Thomas Hart Benton (MMA 2012) and Arthur Dove (Wimsat 1982; Mayer and Myers 2013, 202, 219) and New York City colleagues Richard Pousette-Dart (D. Miller 1990) and Ellsworth Kelly (Wijnberg et al. 2011).[136] Hofmann may have referenced his frequent use of this material in a loose note page among his papers entitled "the might of white" (Hofmann papers, [ca. 1904]–78).

The study group paintings and palettes from the last year of Hofmann's life provide an incomplete picture of the role of recent paint formulations in the artist's late work. Hofmann's late-career paintings incorporate materials embraced by modernists toward the end of his career—such as the phthalocyanine-based blues and greens popular among second-generation Abstract Expressionists including Richard Diebenkorn (Chipp 1957, 46)—but while such contemporary pigments appear regularly in works produced during the last decade of Hofmann's life, it is difficult to ascertain the artist's relationship to the very late additions to his palette.[137] Some colors appear so late in Hofmann's work that it is impossible to determine their impact on the artist. The rhodamine-based magenta paints identified on Hofmann's studio palettes, for example, may reflect the popular interest in rose-colored alizarin alternatives shown by colleagues including Stuart Davis (Seckler 1953, 74) and John Ferren (Campbell 1954, 35),[138] but the absence of rhodamine pigment in any of the study group paintings leaves the question of its overall importance open to debate. The absence of alizarin pigment in study group paintings from the last year of Hofmann's life makes it tempting to associate rhodamine with the "rose-red lake[s]" observed by Lawrence Dinnean in Hofmann's late works (Rockwell 1982, 16), but without definitive identification of this pigment in any study group paintings, no hypotheses can be made at this time regarding the role of rhodamine pigments in Hofmann's compositions. The late appearance of such materials suggests their consideration by Hofmann but does not imply a position among the artist's chosen materials.

Hofmann's palette was gathered from his personal modernist time line, just as his compositions synthesized and advanced formative innovations in late nineteenth- and early twentieth-century painting. Hofmann's materials did not direct his message, but they were intimately tied to the evolving vocabulary of progressive painting and were therefore critical to his modernist vision.

Hofmann's Manipulation of Materials

Hofmann's devotion to a primarily oil-based palette and a limited range of color led to unprecedented explorations of abstract composition. "We revere Hofmann," Minimalist artist Frank Stella later wrote, "for proving that the straightforward manipulation of pigment can create exalted art" (1999, 14). Each component of Hofmann's late-career palette was necessary to his signature style, just as the materials used in each painting appear focused on the task at hand. Through a limited palette, Hofmann's mastery of the essential was carried from means to message. "An idea can only be materialized with the help of a medium of expression," Hofmann asserted, "the inherent qualities of which must surely be sensed and understood in order to become the carrier of an idea" (Hofmann 1948, 64).

Limited Color

The colors identified in the study group paintings advance the exploration of color relationships that defined Hofmann's career. While a wide range of colors were incorporated into early works including *Ecstasy* and *The Third Hand* (both 1947), Hofmann explored an equal array of color relationships in later works, such as *Combinable Wall I and II* (1961; see fig. 3.4), using only a limited number of pigments (fig. 6.2). As noted above, scientific analysis of primary, secondary, and tertiary shades of the red, yellow, blue, and green paints in *Combinable Wall I and II* reveals a palette built largely on three pigment families—cadmium, phthalocyanine, and ultramarine—modified by manufacturers through the addition of semitransparent pigment extenders (see sections on SEM-EDS and XRD in chap. 5). The multiple colors perceived by the viewer in *Combinable Wall I and II* are the effect of neighboring colors, a phenomenon much like the tonal effects described two years later by Josef Albers in his 1963 teaching manual *The Interaction of Color*. Hofmann's use of color, however, goes beyond the interactions observed and catalogued in his fellow teacher's "laboratory course" (Albers 1975, 69).[139] The dimensional appearance of the colored rectangles in *Combinable Wall I and II* is created through Hofmann's activation of color interactions to energize the picture plane. "It is not the form that dictates the colour," explained Hofmann. "It is the colour that makes the form" (Hofmann papers, [ca. 1904]–78).

A different example in Hofmann's exploration of color appears in the 1964 painting *Silent Night* (fig. 6.3). Scientific analysis of the materials in *Silent Night* identified the largest number of disparate pigments utilized by Hofmann in any of the study group paintings, yet the composition is an exploration of primarily two colors: yellow and blue.

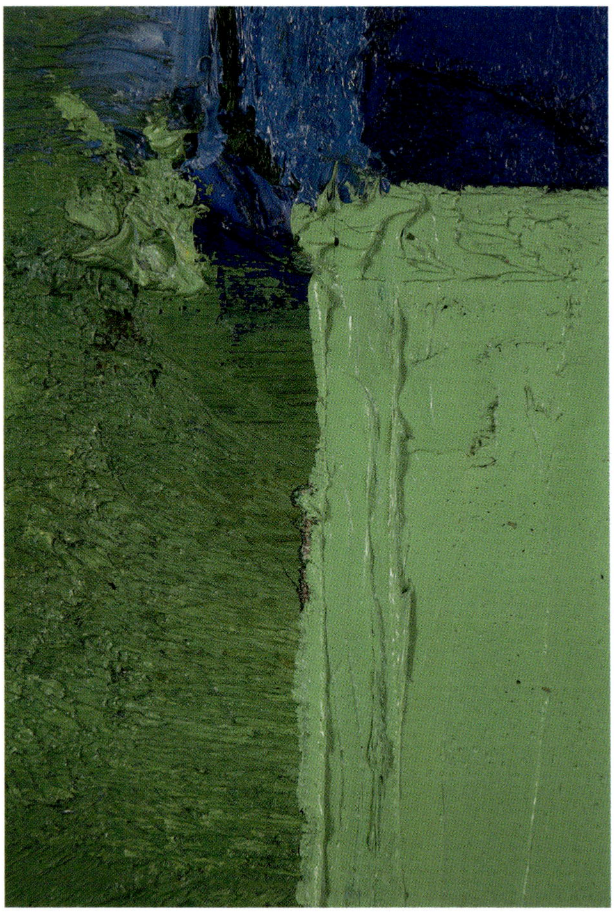

FIGURES 6.2 AND 6.3

Figure 6.2 (left): Detail showing different extenders mixed with similar pigments in Hans Hofmann, *Combinable Wall I and II* (1961; see fig. 3.4). Figure 6.3 (right): Hans Hofmann, *Silent Night*, 1964. Oil on canvas, 213.3 × 198.8 cm (84 × 78¼ in.). BMA/PFA, 1965.5. Gift of Hans Hofmann.

These images illustrate the effects Hofmann achieved through the use of the same pigment modified with semitransparent extenders (left) and a palette exploring primarily two colors through the use of a wide variety of pigments (right).

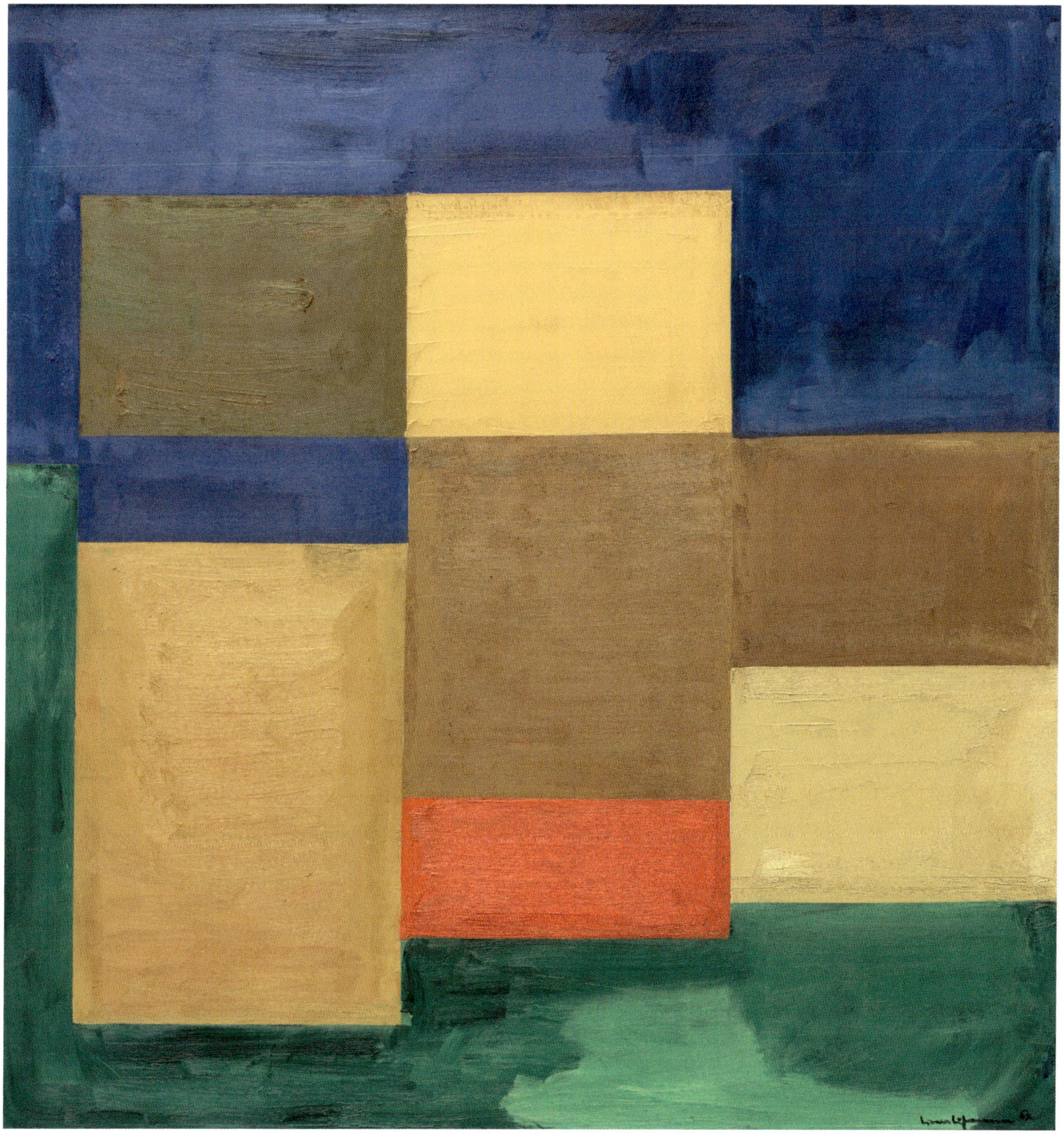

In the progression from the shaded colors of *Combinable Wall I and II* (1961) to the bold hues of *Silent Night,* Hofmann's study of the relationships among colors shifts to the study of relationships within color: "Continuity of color development is achieved through successful, successive development of the color scales.... [and] the rhythmic development of the red scale differs from that of the blue scale, of the yellow scale, etc.... Whereas in tonal painting neighborhood relations are achieved through dark-and-light transitions, in pure painting [there is] rhythmic interweaving of the color scale" (Hofmann 1955, 2).

The geometric forms of both *Combinable Wall I and II* and *Silent Night* showcase the important role Hofmann's controlled color plays in his exploration of the picture plane.

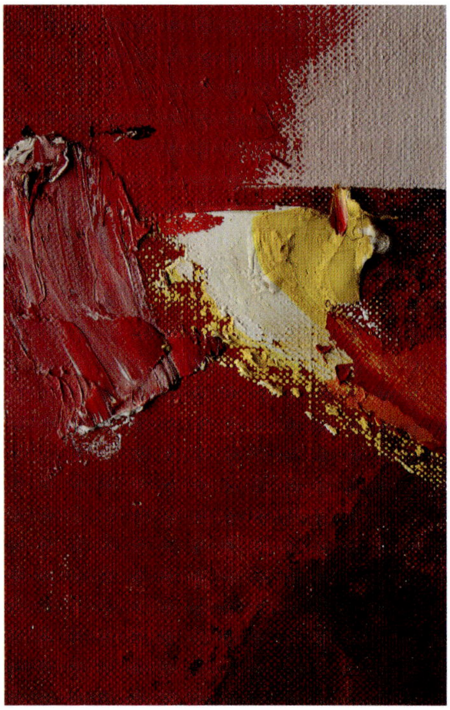

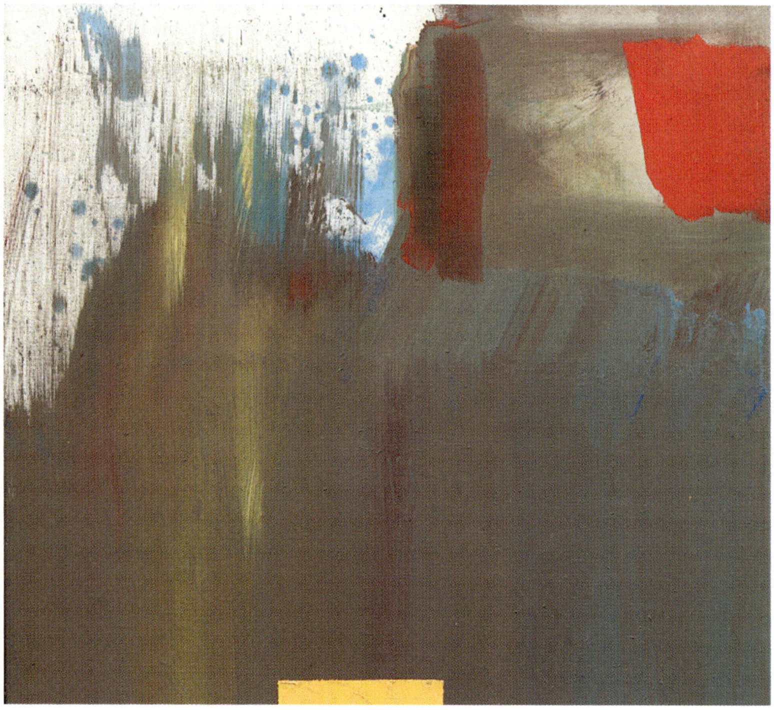

FIGURES 6.4 AND 6.5

Figure 6.4 (left): Detail of Hans Hofmann, *The Clash* (1964; see fig. 3.11). Figure 6.5 (right): Detail of mixed/wiped paint in Hans Hofmann, *Memoria in Aeternum* (1962; see fig. 3.5).

Hofmann's mixed paints generally display their constituent colors, whether the paint is applied heavily (left) or in thin washes (right).

"To create with pure color should be your aim," Hofmann told students attending his Winter 1939–40 lectures. "The color should not be an ornament. The color should be a creative element of the painting" (Kiesler papers, [ca. 1910]–2003, "Hofmann Lecture No. VI"). Hofmann limited, but did not discount, the use of mixed color in his late-career paintings. "Pure color means not at all that the color should be used unbroken or unmixed," Hofmann told his students. "Pure color means only that the colors are properly related[,] that they give a power to each other that they have not in themselves" (Kiesler papers, [ca. 1910]–2003, "Lecture V," 12). Nonetheless, Hofmann's thicker mixed paints generally display their constituent colors (fig. 6.4), as do the streaked washes in works such as *Memoria in Aeternum* (1962; fig. 6.5).

Streaked color blending may be the result of Hofmann dragging a wet, paint-soaked piece of gauze across the canvas, a practice described by Elaine de Kooning in 1950 (58–59). According to studio assistant Wolf Kahn, muddied tones were also manufactured from a day's unsuccessful paintings. Kahn recalled that Hofmann often scraped wet paint from discarded compositions and saved it in wax paper for later use. "That's my favorite color," Hofmann told Kahn, "because I use it as a contrast against bright new color.... I call that the *Scheisse*" (Dickey 2011, 243).[140]

A less overt use of mixed color appears in a group of paintings Hofmann produced between 1959 and 1964, including *Ruby Gold* and *Indian Summer* (both 1959; see figs. 3.6 and 3.7), *Combinable Wall I and II*, and *Silent Night*. These paintings are distinguished from Hofmann's other works of the same period not only by strongly geometric compositions but also by what appears to be the related use of localized preparatory layers. In these few works, warm colors are painted over a ground layer of white alkyd paint, while cool colors are applied atop a layer of mixed black and phthalocyanine green oil paints brushed directly onto a bare canvas. In regions where warm and cool colors meet, the ground layer alternates between alkyd- and oil-based materials (fig. 6.6). In addition, there is no ground application beneath some orange paints in *Ruby Gold* and *Indian Summer*; this creates localized voids in the patchwork ground application.[141] Like

FIGURE 6.6
Detail of localized black and white ground layers in Hans Hofmann, *Combinable Wall I and II* (1961; see fig. 3.4), which exemplifies the alternating ground colors found in a handful of paintings produced by Hofmann between 1959 and 1964. Warm colors are painted over a layer of white alkyd paint, while cool colors are applied over a layer of mixed black and phthalocyanine green oil paints.

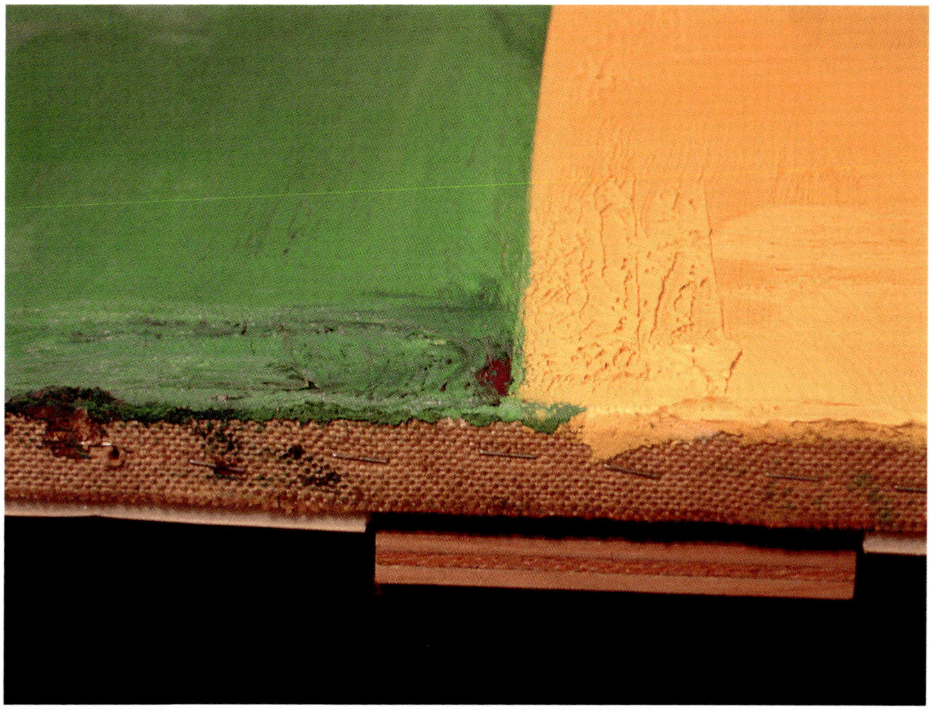

Hofmann's late-career switch to alkyd ground and priming materials, this period of experimental ground layer application has direct correlations to condition issues in the artist's work, as this work will endeavor to demonstrate.

Limited Paint Media

Hofmann's use of paints made with new binding media was limited. As noted previously, only alkyd paints appear alongside the oil-based artists' tube colors utilized in Hofmann's late-career paintings. Despite Hofmann's wholesale adoption of alkyd ground layers in the 1950s, the appearance of these industrial paints in his compositional layers is restricted to a small number of works produced over a three-year period. Commercial house paints were commonly used as priming materials by Abstract Expressionist painters such as Jackson Pollock, Franz Kline, and Barnett Newman (Rogala et al. 2010; Wijnberg et al. 2011), and colored industrial alkyds played a prominent role in the drip paintings of contemporary artists including Janet Sobel and Pollock, who, according to conservator Susan Lake, was "taking full advantage of the relatively new alkyd-resin paints by 1949" (Lake, Ordonez, and Schilling 2004, 140). Given the availability of and interest in industrial paint media among Hofmann's colleagues and given Hofmann's adoption of modernist color, the relative scarcity of alkyd paint in his compositions is likely an intentional restriction.

Only three of the study group paintings include alkyd paint in their compositional layers. The alkyd paint appears in the form of a black splash, in either the foreground or the background. Four of the study group paintings include areas of splashed black oil paint: *The Vanquished* (1959), *Bald Eagle* (1960), *Tormented Bull* (1961), and *Heraldic Call* (1962). Analysis of materials from these paintings shows that the black splash of *The Vanquished* is composed entirely of oil paint, while splashes on the remaining paintings include both oil and alkyd paints. Just as Hofmann's white alkyd grounds do not preclude his use of white oil paint—zinc white oil paint appears in the compositional layers of all the study group paintings except *Heraldic Call*—Hofmann's use of both alkyd and oil black paints is not arbitrary. These mixed-media splashes are artifacts of transition,

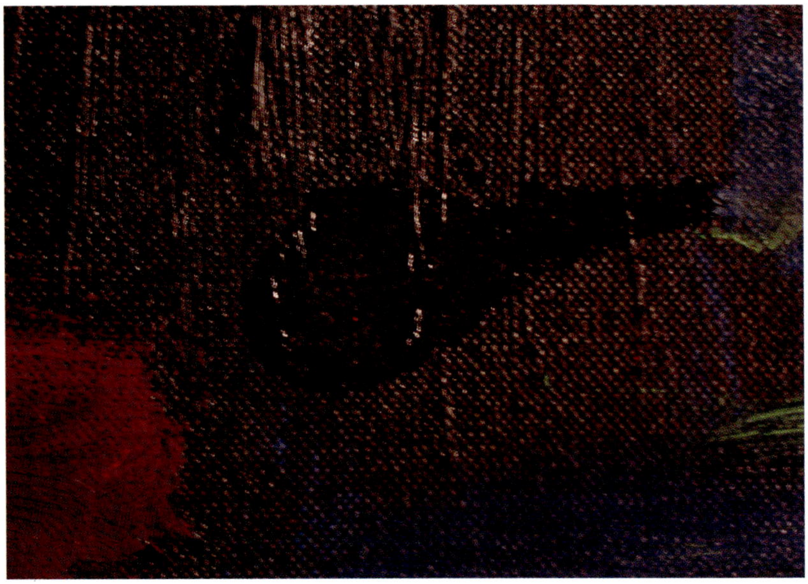

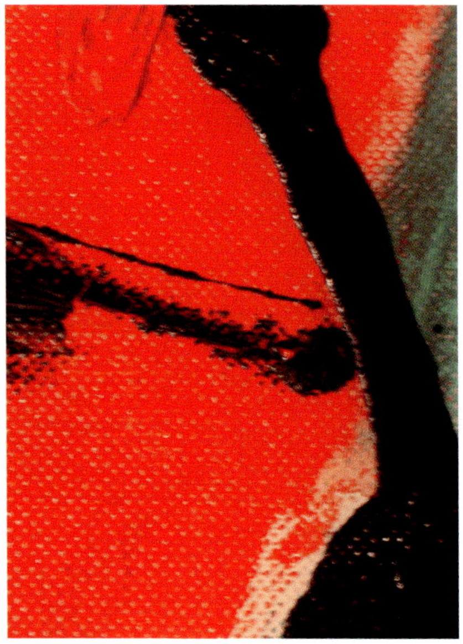

FIGURES 6.7 AND 6.8
Details of dripped alkyd paint over (left) and next to (right) brushed black paint in Hans Hofmann, *Bald Eagle*, 1960. Oil on canvas, 153 × 132.7 cm (60¼ × 52¼ in.). BAM/PFA, 1964.3. Gift of Hans Hofmann.

Thinned applications of oil paint lack the color intensity and the surface characteristics of their alkyd counterparts.

evidence of Hofmann's attempts to adapt the effects of another paint material to the conditions imposed by his oil-based palette. The black splash of *The Vanquished* (see fig. 3.10) is an extension of the splashed ink noted by Rockwell in the artist's earlier work (1982, 4), but the thick, wrinkled surface of the puddled oil paint (see fig. 4.10) cannot mimic ink's flat, opaque surface. Thinned applications of black oil paint in Hofmann's subsequent "splash" paintings are no substitute for the color intensity and the surface characteristics of their alkyd counterparts (figs. 6.7 and 6.8).

Black paint—both oil and alkyd—disappears from the background of the study group paintings after 1962, replaced by the rich alizarin washes and splashes of *Magnum Opus* (1962), *Polyhymnia* (1963), and *The Clash* (1964; see fig. 3.11). Black played only a limited role in Hofmann's late-career compositions outside of the splash experiments, and it disappeared entirely from the artist's palette shortly thereafter. No black pigments appear in any of the study group paintings after 1963. In abandoning the use of both alkyd paints and black pigments from his compositional layers in the late paintings, Hofmann regained the vibrant, bold palette at the core of his work.

Handling characteristics may also have played a role in limiting Hofmann's use of new paint media. Hofmann's physical painting style relied on the resistance of a solid material, "address[ing] the picture surface as a responsive rather than inert object," recalled Greenberg, "and painting itself as an affair of prodding and pushing, scoring and marking, rather than simply inscribing or covering" (1961, 24). Alkyd paints spread out into thin films and dried quickly, characteristics ill suited to the oil painting technique of Hofmann and colleagues such as Robert Motherwell, who found his traditional painting practices challenged by the new paint media: "If you are working with a half a dozen buckets of liquid—like house paint on the floor, the options of mixing not only are less seductive, in a way they are more a pain in the neck...the liquidity to begin with means that one can't be involved in nuances of thickness or manipulating it with one's hands....It becomes much more artificial to nuance with buckets of paint" (Caws 2003, 114).

Instead of adapting his practices to accommodate the new paint media, Hofmann stretched the boundaries of his established palette to explore contemporary painting styles. No mere reflection of modern method, Hofmann's adaptation of advancing art practice advanced his own painting technique. The alizarin-splashed backgrounds of *Magnum Opus* and *Polyhymnia* (figs. 6.9 and 6.10), for example, show close allegiance with the stained canvases of Mark Rothko and Helen Frankenthaler from the same period. "When he made those very turpentiney paintings," recalled Frankenthaler, "it was as if he were saying, 'I'll fix you, you kid. I'll show you how it's really done'" (Brown 1998, 31). Hofmann's skilled manipulation of paint allowed him to explore contemporary style while simultaneously developing his mastery over a dedicated palette of materials.

Hofmann's late-career palette was the springboard from which the artist achieved his most innovative compositions. By restricting his materials, Hofmann delved deeper into the essential components of painting. "Plastic creation asks for feeling into the essentuality [*sic*] of nature as well as for feeling into the essentuality of the nature of the medium of expression," wrote Hofmann in a statement dated May 1, 1951. "The plastic experience gained by the former must be transformed into the plastic language of the other" (Hofmann papers, [ca. 1904]–78). In a 1964 compilation of essays entitled *The Anxious Object*—reprints from 1966 onward include a dedication to Hofmann—Harold Rosenberg ruminated on the rewards of focused practice: "Instead of solving his problem—'his' because he has chosen it—the artist lives it through the instrumentality of his materials. By fixing his idea in matter... he is led to experiment and refinement. In time, he becomes so adept in materializing his hypotheses, and in manipulating his materials as if they were meanings, that the problem itself is transformed" (Rosenberg 1982, 22).

In his art practice Hofmann melded medium and message, and in doing so, the artist foreshadowed a dialogue central to subsequent modern art practice. "Hofmann somehow recognized the advanced, process-driven art of the early sixties," Frank Stella noted. "[He] helped make the greatest change in twentieth-century art both possible and successful" (1999, 16–17). It is no coincidence that the first steps toward T. J. Clark's "In Defense of Abstract Expressionism" (1994) were taken at a lecture in front of the Hofmann Collection at the University of California, Berkeley. "What matters is Abstract Expressionist paint is once again a medium," Clark told the audience. "What matters is the ability of this painting to retrieve the process of painting, and its bare material condition as a fact worth painting" (Clark 1992).[142]

The Correlation Between Condition Issues and Hofmann's Use of New Materials

Understanding Hofmann's use of materials is also the key to preserving his legacy. While his late-career works exhibit many condition issues common to other oil paintings on canvas, there are a handful of instances where condition issues arise in combination with specific materials in his palette. These issues appear in conjunction with Hofmann's use of new, mid-twentieth-century paint media and pigment formulations. An unexpected incompatibility of commonly used materials and methods is responsible for the recurring examples of lifting paint, canvas distortion, and efflorescing paint layers in Hofmann's late-career paintings.

White Paint and Ground Materials

Hofmann's use of zinc white oil paints and alkyd house paint ground layers has directly impacted the condition of his late-career paintings. The alternating heavy paint layers, underbound washes, and splashed forms of Hofmann's signature works place unusual stress on the white paint and ground layer materials favored by Hofmann and his colleagues.

Weak pigment matrices and engineered failure mechanisms inherent in the bright zinc-based white paints favored by mid-twentieth-century artists made these materials ill suited for use as fine art paints. My previous research—performed in collaboration with conservators and scientists from the Smithsonian Institution and the National

FIGURES 6.9 AND 6.10

Figure 6.9 (left): Hans Hofmann, *Magnum Opus*, 1962. Oil on canvas, 213.7 × 198.4 cm (84⅛ × 78⅛ in.). BAM/PFA, 1963.7. Gift of Hans Hofmann. Figure 6.10 (right): Hans Hofmann, *Polyhymnia*, 1963. Oil on canvas, 183.2 × 153 cm (72⅛ × 60¼ in.). BAM/PFA, 1964.1. Gift of Hans Hofmann.

These images illustrate how Hofmann explored the effect of the acrylic-stained unprimed canvases of Mark Rothko and Helen Frankenthaler while retaining his dedication to bright white ground layers and bold oil-based paints.

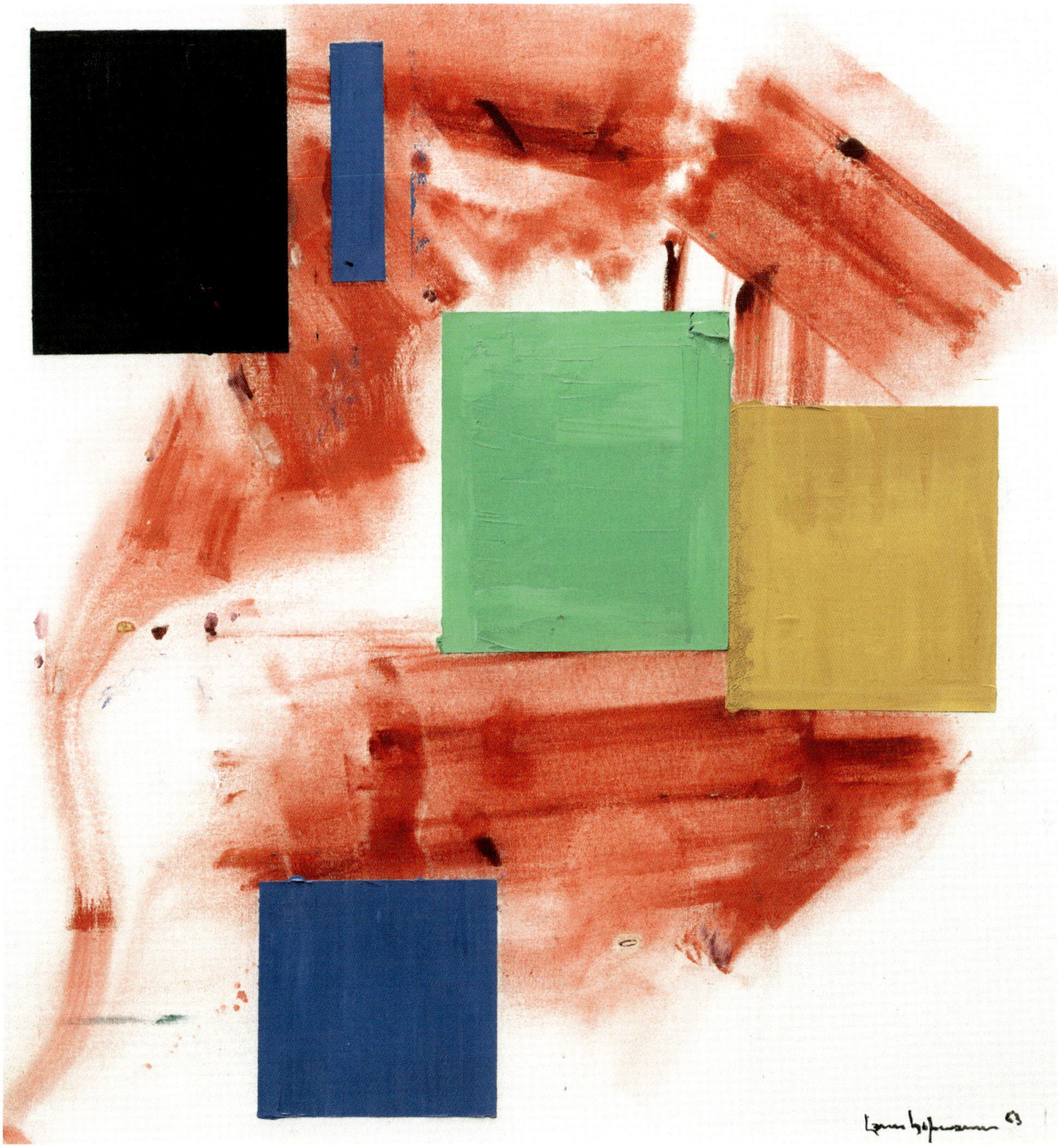

Gallery of Art (Rogala et al. 2010; Maines et al. 2011) examined the drying mechanism of zinc oxide oil paints and determined that their stiff and brittle qualities do not suit them for the flexibility and load-bearing requirements of Abstract Expressionist paintings. Zinc oxide attains a lamellar distribution in oil medium wherein layers of fatty acid chains lie between layers of fatty acid carboxyl groups and the zinc matrix. This layered structure makes the paint very stiff and difficult to oxidize. Consequently, unsaturated fatty acids in zinc oxide paint remain trapped within the paint layer years after oxidation of the paint should have been completed. The paint has prematurely "frozen" into position, without achieving the structural stability afforded by the cross-linking that would accompany the natural drying process. Rather than constituting a well-formed paint layer made up of a uniform cross-linked network, a paint film containing primarily zinc oxide pigment consists of a collection of plate-like layered "islands" prone to separation and internal (intralayer) cleavage of the paint layer.

Problems inherent in paintings with zinc oxide ground layers were exacerbated in the mid-twentieth century by the appearance of ineffective house paint formulations of zinc oxide marketed during the period when Hofmann and his colleagues experimented with house paint grounds. Early house paints, such as those used by Hofmann and his colleagues, were not formulated for long-term stability. Frequent statements in the contemporaneous literature confirm that while zinc oxide was considered a poor performer in terms of film formation, the industry requirements for durability differed significantly from those of the modern artist. "A white house paint possesses real merit," noted one New Jersey Zinc Company representative, "if it maintains a...surface free of significant film failure for a period of three years" (Werthan 1947, 38). Early house paint manufacturers were also aware of the problems posed by layering zinc oxide house paint with other paint materials. "In the aim of developing a better primer [a paint used as a foundation for subsequent paint layers] it is possible that too little thought has been given as to how this primer might work under the different finishing paints," another company representative noted in a 1935 article for *Paint, Oil, and Chemical Review*. "In some cases there is a marked increase in checking and cracking of the finishing coats and in others an actual decrease in adherence of the whole system" (Schmutz 1935, 356). This type of historical literature sheds light on the widespread condition problems observed in the zinc oxide paint layers of modern paintings. In the 2010 study (Rogala et al. 2010), all eighteen of the Abstract Expressionist paintings that exhibited severe paint layer failure were found to have zinc oxide grounds or underlying layers of compositional zinc oxide paint, and all of the paintings in the current study group with zinc oxide paint layers have exhibited cracking and cleaving paint (see table 2). Historical literature on zinc oxide has only recently entered the conservation research canon (Rogala 2014; Osmond 2012).

The white alkyd house paint formulations also popular with Hofmann and his Abstract Expressionist colleagues avoid the problems associated with zinc oxide pigment but exhibit other characteristics ill suited to fine art materials. In the mid-1950s, heavily bulked mixes of titanium dioxide pigment replaced zinc oxide in American house paint formulations. The transparency of titanium dioxide in oil binders had previously relegated the inexpensive pigment to use as an extender, or filler, material in artists' oil paints, but titanium dioxide alkyd formulations provided industrial paint manufacturers with economical house paints less vulnerable to the wide-ranging climate conditions of the American market. Titanium dioxide alkyd paint exhibits failure behaviors similar to those found in zinc oxide oil paints. The high pigment-volume concentration endemic to house paint formulations (Feller 1964, 25–26) creates a weak paint film whose weakness is exacerbated by the brittle nature of an alkyd paint binder. Alkyd paint films will fail before oil paint films under equal amounts of stress, and in layered structures of oil and alkyd paints, the stress imposed on an alkyd film will transfer to surrounding materials upon failure of the alkyd paint layer (Erlebacher et al. 1992; Hagan et al. 2007). Mechanical stresses on paintings with disparate paint layers can result in widespread failure of the composite structure when the stress of failed paint layers is transferred to equally vulnerable paint films (Young et al. 2004; Maor and Murray 2008). An example of this phenomenon in the study group paintings is the early and ongoing cracking of layered zinc and alkyd priming materials in the painting for which Hofmann signed the insurance waiver, the 1961 work *Tormented Bull* (figs. 6.11 and 6.12).

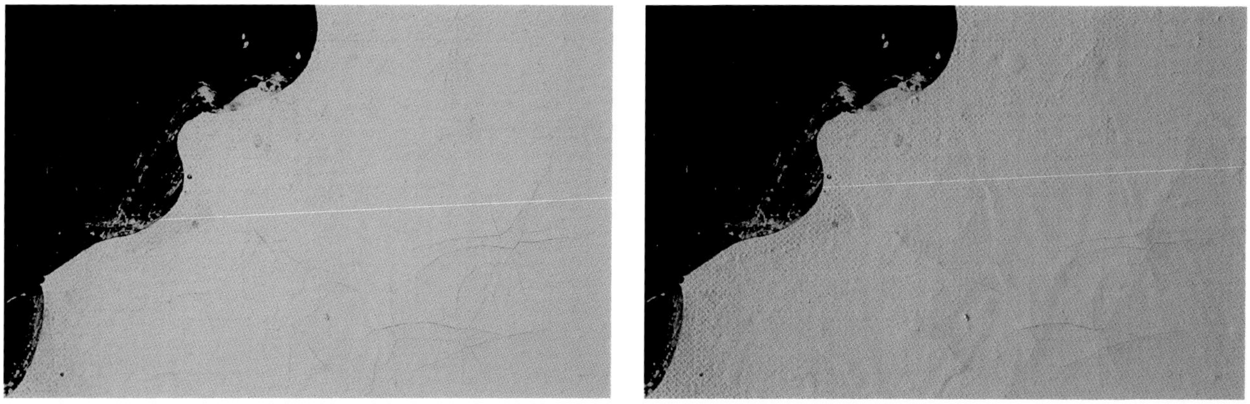

FIGURES 6.11 AND 6.12
Detail of a cracking priming layer in Hans Hofmann, *Tormented Bull* (1961; see fig. 3.8), photographed in normal (left) and oblique (right) light. Oblique light highlights the widespread cracking of Hofmann's alkyd priming layer, painted over a commercially applied ground; the cracking of alkyd ground layers is usually obscured by compositional paint.

What is significant about problems in the ground layers and compositional white paints used by Hofmann and his colleagues is that they are the unexpected result of combining Abstract Expressionist painting method with the painting materials commonly used by these artists. Zinc oxide oil paint and alkyd house paint were not engineered for the load-bearing function historically fulfilled by lead white artists' paints, but these new paints were popular choices from among limited alternatives as lead paint left commercial manufacture. The problems inherent in Hofmann's choice of preparatory materials are compounded in those instances where Hofmann used multiple ground layer materials. The different drying and absorption characteristics of Hofmann's oil and alkyd paints directly contribute to the localized distortions present in works such as *Ruby Gold*, *Indian Summer*, *Combinable Wall I and II*, and *Silent Night*. Differences in the expansion-contraction process during oxidative drying of the paints can distort the picture plane and place additional stress on individual paints and on bonds among paint layers. Localized binder loss in compositional paints that span both saturated (oil) and absorbent (alkyd) grounds can also physically distort the paint layers through uneven drying. Localized binder loss may also have been exacerbated by the lack of glue sizing in Hofmann's late paintings, although accounts of sizing application in Hofmann's practice are inconclusive. While Erle Loran states that Hofmann always sized his canvases, and while ultraviolet examination by Carolyn Tallent identified fluorescence consistent with glue-based sizing materials on two early study group paintings (1987; 1988, 6), an account by former San Francisco Museum of Modern Art conservator Will Shank states that by the mid-1950s, Hofmann's alkyd ground layers replaced earlier glue sizing and gesso preparatory methods (1991, 3).[143] Gaps in alkyd paint left by alternating ground preparation therefore would allow direct contact between oil paint and canvas, leaching binder from the paint layer and exposing the canvas to acidic paint materials. Differences in drying behavior and differences in topography between structural oil paint and conforming alkyd paint surfaces place increased stress on the borders where oil and alkyd regions meet and on the canvas and upper paint layers that lie across these borders. The distortions endemic to these paintings are exaggerated as the materials age. The problems of loosening fabric and hardening paint are common to all works on canvas, but the effects of aging are exaggerated in works containing alternating ground layers. The uneven stresses within these paintings increase with the disparate aging of their materials and with those materials' differing capacities to withstand stress. While many of Hofmann's late-career paintings exhibit canvas distortion, distortions are particularly pronounced in works with alternating ground layers. Voids in the alternating ground layers in *Ruby Gold* (figs. 6.13 and 6.14) and *Indian Summer* (see figs. 4.2 and 4.3) produce the most extreme distortions seen in the study group.

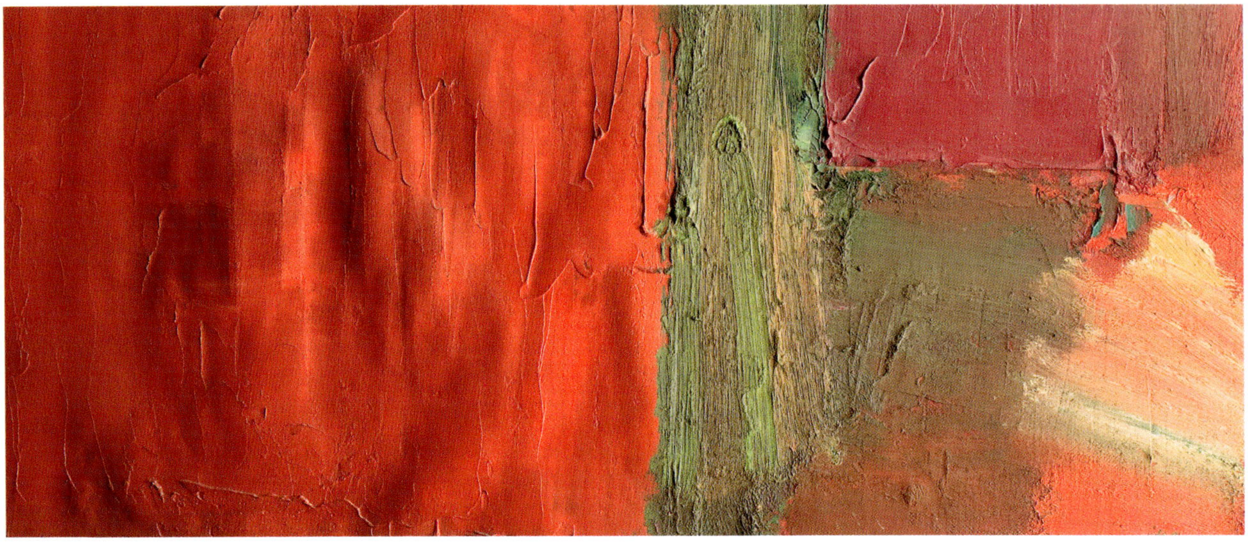

Degradation of Synthetic Colors

Color-specific surface exudates are another notable example of condition problems resulting from a combination of new paint materials and traditional painting technique in Hofmann's work. Unusual surface drying problems, ranging from efflorescing fatty acids to expressed liquid oils, have been observed in relation to a specific formulation of alizarin in every study group painting in which that pigment appears. In contrast to dry exudates found in association with other paint colors in Hofmann's late-career work, this particular alizarin paint exudes both dry and liquid exudates decades after the painting's completion. The pigment was identified through analysis as Pigment Red 83 (see chap. 5, "Fourier Transform Infrared Spectroscopy"), described in Herbst and Hunger's *Industrial Organic Pigments* as an American-manufactured substitute for natural madder pigment (2004, 9).

Conservation studies have linked chronic efflorescence in alizarin colors with the early exposure of paintings to high levels of humidity, similar to those found at Hofmann's Provincetown studio (Rimer et al. 1999), but there may also exist a relationship between Hofmann's use of PR83 and his painting technique. According to Herbst and Hunger, PR83's alkaline-based formulation is overly responsive to "common organic solvents" (2004, 511), a category that may include the benzine (petroleum ether) solvent Hofmann mixed with his paints to speed their drying in the coastal environment of Provincetown. Traditional painting manuals commonly recommended benzine as a diluent, particularly when paintings were created in situations that slowed the drying of a paint film. "If painting be done in an atmosphere where the humidity is high and the temperature near the dew point," wrote New York City chemist and professor Maximilian Toch in the 1925 manual *The Chemistry and Technology of Paints*, "a great advantage is to be obtained by the moderate use of benzine...in brushing on a quick-drying paint containing benzine the evaporation carries with it much of the moisture in the paint" (Toch 1925, 269).[144]

FIGURES 6.13 AND 6.14
Detail (left) and overall view (right) of Hans Hofmann's *Ruby Gold* (1959; see fig. 3.6), photographed in oblique light. MAG, 60.37. Marion Stratton Gould Fund.

These images illustrate local distortions caused by voids in alternating ground layer materials (see also figs. 4.2 and 4.3).

As noted previously, there is evidence that Hofmann purchased benzine for his classroom and used the material in his painting practice. "When he uses thin paint," de Kooning noted, "[Hofmann] tilt[s] the canvas one way or another to control the 'runners'" (1950, 58). There is also evidence that Hofmann used benzine when painting outdoors. In a series of letters written to student Alice Hodges in the 1940s, Hofmann noted the oppressive weight of his painting supplies. "I can work only in the closest neighborhood," Hofmann complained to Hodges in a July 29, 1943, letter from Provincetown. "It is the

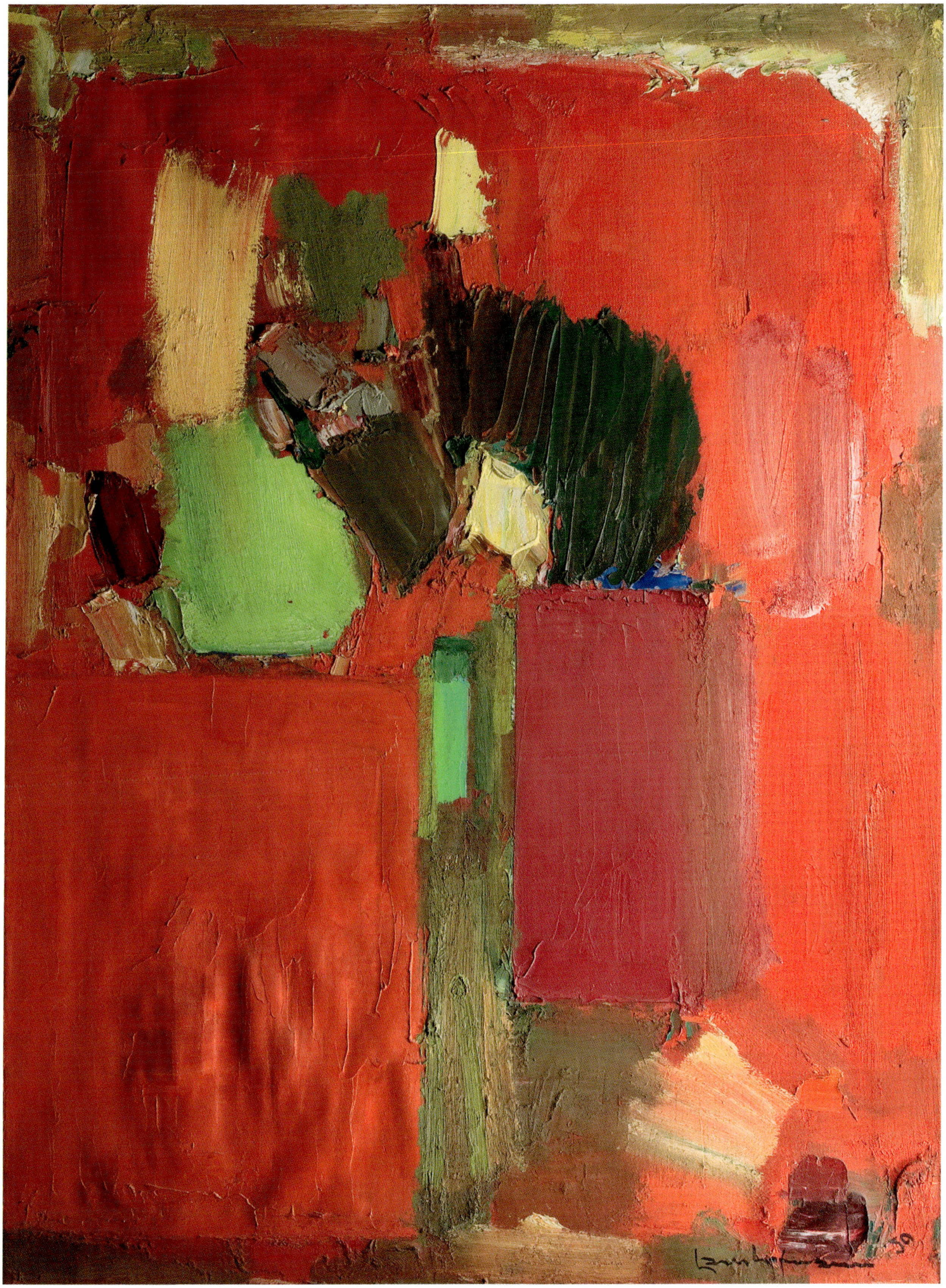

oil painting material [that] is so heavy to carry" (Kiesler papers, [ca. 1910]–2003). Hofmann tethered large glass bottles to his easel when painting on the dunes of Cape Cod and may have carried solvent in those bottles (fig. 6.15). It is possible that Hofmann filled empty bottles with seawater to make his ballast, but the burden Hofmann recounts seems unlikely with empty bottles. "I [have been] only 3 time[s] on the beach," Hofmann wrote Hodges on September 13, 1943, as the summer season neared its end. "The worst is carrying 70–90 pound[s of] material often twice a day over hills up and down."

Conservation and archival accounts of the study group paintings report increased levels of exudation in heavily thinned applications of PR83 in the study group paintings, a finding that is in keeping with new research on the behavior of modern synthetic pigments. Research presented at the 2013 symposium "Issues in Contemporary Oil Paint" linked the presence of exudates in modern oil paints with the separation of binding media and pigment in response to added solvents—with attendant polar agglomeration of the pigment particles (Boon and Hoogland 2014; Heydenreich et al. 2014)—and tests of solvent sensitivity in synthetic organic pigments categorized PR83 as the most reactive and sensitive pigment of the study group (Blumenroth et al. 2014). If a connection exists between Hofmann's use of benzine and condition issues with his late-career alizarin paints, the phenomenon may again be the result of incompatibility between traditional painting practices and newly formulated painting materials. As in Hofmann's use of house paint grounds, the problem of exudates in his alizarin paint colors may arise from the expected incompatibility of favored material and method. Hazardous overlaps in accepted practice may also arise in the conservation treatment of these alizarin exudates, which were commonly removed by cleaning with benzine, a conservation solvent considered safe for the testing and removal of exudates (SFMOMA records 1966–present).[145]

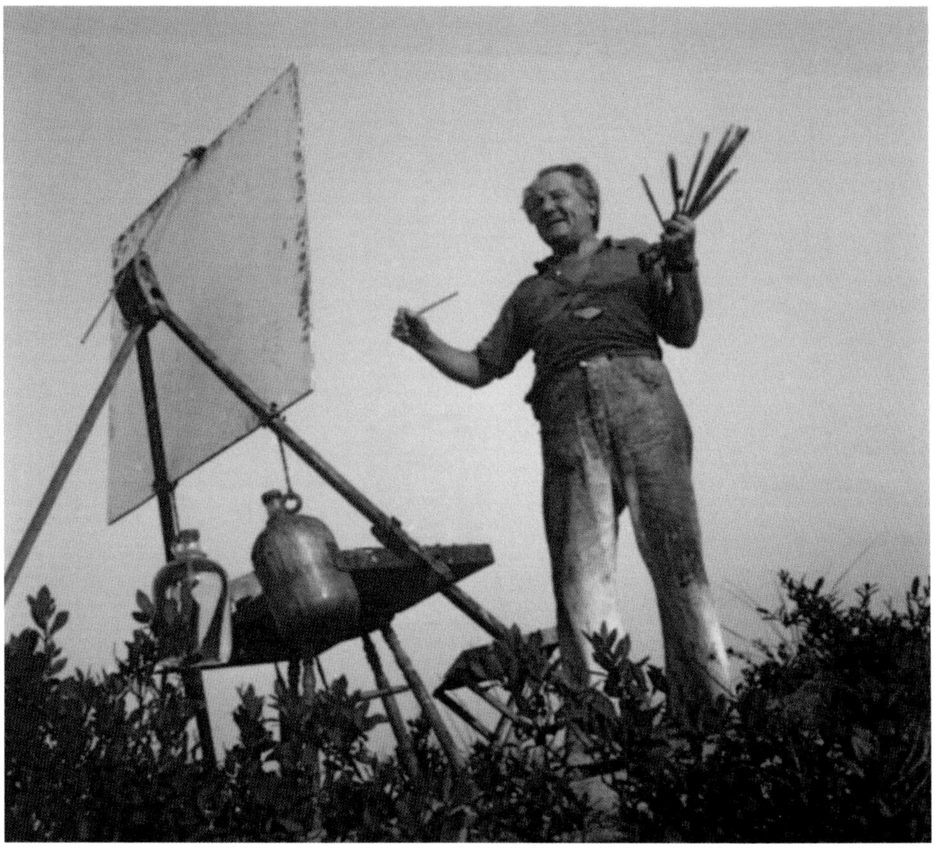

FIGURE 6.15
Hofmann painting on the dunes of Cape Cod, 1943. Note the bottles of liquid tied to the easel.

Photo: Herbert Matter.

Conclusion: The Lessons of Hofmann's Late-Career Materials

We can call the artist's technical procedure a self-fashioning, a making of self, since it produces or defines the artist as artist as much as it produces the artwork.
—RICHARD SHIFF, "Performing an Appearance: On the Surface of Abstract Expressionism," 1987

The medium must not be separated from the content which it embraces.
—WILLIAM CHAPIN SEITZ, "Abstract-Expressionist Painting in America," 1955

The work of renowned Abstract Expressionist painter and teacher Hans Hofmann shows us why and how we need to alter our thinking about the study and conservation of modern art. The research presented in the preceding chapters aids our efforts to preserve Hofmann's physical legacy and expands our understanding of Hofmann's role in the modernist communities of Europe and the United States. In positioning Hofmann as part of a modernist continuum, I wish to place modernist materials within the same framework and redirect our approach to modern art scholarship.

Hofmann's participation in the formative modernist communities of Munich and Paris afforded the artist a central position within the American avant-garde when he arrived in the United States in 1930, and it placed Hofmann's schools in New York City and Provincetown, Massachusetts, at a primary nexus of experimental painting theory and practice. Hofmann's influence with colleagues, critics, and the thousands of students and teachers who flocked to his schools—"as a pilgrim comes to Mecca," recalled student Nell Blaine (Hofmann Student Dossier 1963)—made the Abstract Expressionist painter and teacher simultaneously the influencer and the benefactor of the leading edge of modern American art. In writing on the "present prospects of American painting and sculpture" in 1947, critic Clement Greenberg asserted that "Most of the young artists [today] have either been students of Hans Hofmann or come in close contact with his students and ideas.... Hofmann will in the future, when the accomplishment of American painting in the last five and the next twenty years is properly evaluated, be considered the most important figure in American art of the period since 1935 and one of the most influential forces in its entire history" (Greenberg 1947, 29).

The materials and techniques employed in Hofmann's paintings represent a key moment in the evolution of modern art—a paradigm shift from product to process that influenced subsequent generations of artists. The overt materiality of Hofmann's signature compositions can be seen as a direct link between the representative, nature-based abstractions of the late nineteenth- and early twentieth-century modernists and the conceptual, materials-based process art that rose to prominence after the Second World War. The study of Hofmann's late paintings underscores the formative role that materials studies play in interdisciplinary scholarship and reveals how the study of physical evidence in historical context informs our understanding and preservation of modern art.

Hofmann as a Case Study in the Ramifications of Mid-Twentieth-Century Painting Practice

Hofmann's late-career paintings are constructed from a selected range of modernist paint materials that reflect the artist's connection to centers of art and materials innovation in the late nineteenth and early twentieth centuries. Hofmann continued to incorporate new paint materials into his work throughout his career, but notable shifts in the materials used in Hofmann's late paintings from those employed in the artist's earlier work are primarily found in his preparatory and support layers—in Hofmann's change from cotton to linen fabric supports and in his late-career adoption of alkyd house paint ground layers. Analysis of materials from the study group paintings confirms published and personal accounts of the artist's almost exclusive use of oil-based paints for his boldly colored compositions. Deviations from Hofmann's oil-based palette include the limited use of alkyd compositional paints, which appear for a brief period in works created shortly after he retired from teaching. Splashed applications of black alkyd paint appear in Hofmann's works from 1959 to 1962, at which time he replaced the alkyd splashes with brushy applications of alizarin oil color.

The period after Hofmann's retirement is also distinguished by a brief period of experimental ground layer preparation, when a handful of geometric works produced between 1959 and 1964 were executed over localized preparatory layers—regional oil grounds underneath cool colors and alkyd grounds underneath warm colors, with occasional voids in the ground beneath orange colors—a technique that does not appear at any other time in Hofmann's late work. Condition issues related to paint application—including this experimental ground technique—are seen throughout Hofmann's late-career work. In addition, condition issues related directly to individual materials appear consistently in association with the artist's use of zinc white oil paint, alkyd paint, and alizarin oil paint. All three of these modern materials perform poorly in response to modern painting techniques and to the functions assigned to them by the construction of modern paintings. Zinc oxide oil paint and alkyd house paint, for example, were embraced by modern artists as alternatives to toxic lead white paints, but neither material was engineered for the load-bearing function historically fulfilled by lead white paints and required by heavy Abstract Expressionist paint layers. Similarly, synthetic alternatives to natural madder pigment—such as the PR83 alizarin pigment that appears throughout Hofmann's late work—were an improvement over natural madder pigment, prone to fading and discoloration, but the synthetic alternatives respond poorly to the conditions under which the artists painted and to the additives in common use by the same artists. While these materials behaved as they were designed to behave, and while they addressed Hofmann's needs at the time of their use, their inherent physical properties are at odds with the desire of conservators and scholars to keep the works looking as the artist intended.

Hofmann's late works were created at a pivotal moment in the dialogue between artist and art making, but the notion that twentieth-century modernist innovators such as Hofmann abandoned traditional painting practice is a myth that works against the study and preservation of their artworks. The comprehensive catalogue of Hofmann's materials presented in this study reveals a primarily traditional palette, with only limited incorporation of modern paint media. Apart from the artist's adoption of alkyd ground layers, the distinctive visual vocabulary of Hofmann's later years is fashioned largely from traditional artist's oil paints. Many prominent Abstract Expressionist artists employed a similar mix of familiar and experimental materials, such as Willem de Kooning (Hess 1953; Lake, Lomax, and Schilling 1999) and Franz Kline (Rogala et al. 2010), who continued to work with traditional artists' oil paints while experimenting with commercial house paints and industrial alkyd coatings.[146] Published descriptions of visits to the studios of de Kooning and Richard Diebenkorn recount "tube oil colors in random piles" found next to "a variety of half-pint cans of oil-based house paint" (Chipp 1957, 46), and traditional oil paints have also been found in combination with alkyd and acrylic paints in works by Barnett Newman and Ellsworth Kelly (Wijnberg et al. 2011). Ad Reinhardt reportedly had no interest in modern paint materials (Lippard and Hunter 1967, 26), while Jackson Pollock—despite his well-known affinity for industrial paint formulations—continued to incorporate traditional oil paints in his compositions late into his career (Coddington 1999; Lake, Ordonez, and Schilling 2004).

The mix of new paint materials with traditional materials and painting practices that characterizes Abstract Expressionist painting has a direct impact on the material stability and aging characteristics of artworks from this pivotal art movement. Condition issues in Hofmann's late work appear consistently in association with the artist's use of zinc white oil paint, alkyd paint, and alizarin oil paint. In each of these cases, the modern material chosen by Hofmann behaved according to its design but performed poorly in response to commonly used painting techniques and did not meet the structural requirements of paintings that combined new and traditional art materials. Condition issues in Hofmann's work appear in relation to the artist's use of zinc oxide oil paint—a popular, less toxic alternative to lead white that proved too stiff and brittle for Hofmann's heavy compositions and physical painting techniques—and his late-career adoption of ground layers made from alkyd house paints—whose weak pigment matrices and engineered failure mechanisms made them ill suited for use as fine art materials and for layering with traditional oil paints. Similarly, Hofmann's preferred synthetic alizarin pigment responded poorly to the conditions under which the artist painted and to additives in common use by period artists. This study of Hofmann's late-career paintings sheds light on an unexplored incompatibility between the new materials associated with Abstract Expressionist painting and the conventional art materials and practices still used by many Abstract Expressionist artists.

The condition problems in Hofmann's work—which are related to the mix of material and method endemic to Abstract Expressionist painting practice—lead us to reconsider the role of materials in the aging and degradation of modern paintings. The history of modern painting includes the gradual incorporation of new materials, and it is shortsighted to believe that the physical impact of modern paints lies with anomalous failures in paint formulation, or that an artist's use of modern materials warrants study only when those materials play a prominent role in the artist's technique. The works of Hofmann and his Abstract Expressionist contemporaries set as standard art practice a mixture of techniques and materials that directly impacted the physical stability of works by these painters and their modernist descendants. Some modern materials, such as the PR83 synthetic alizarin pigment mentioned above, may respond poorly to traditional conservation treatment—in this case, the use of benzine to test and remove exudates. In other cases, the localized treatment methodologies commonly applied by conservators in response to site-specific condition issues in paintings do not take into account the systemic condition problems—revealed by this research—in such structurally important components as ground layer material. Traditional, localized treatment strategies are not sufficient to address the widespread failure of such materials in paintings from this era —condition problems that can propagate and become more severe in response to changing environmental or handling conditions. Modern paintings and painting collections exhibiting localized condition problems should be assessed for the presence of mixed materials and the attendant potential for less visible, widespread condition concerns; modified treatment methodologies should be considered, such as protocols that consider incorporating overall treatment, the use of overall support materials, and limitations on travel and exhibition schedules.

The problem of addressing overall condition issues, however, is not as simple as implementing overall treatment strategies, as treatment methodologies intended to address widespread condition issues must be balanced against the increasing role of physical evidence in art historical study. The future study of Hofmann's regional ground layer applications, for example, is directly affected by early conservation treatments that eliminated the canvas distortion indicative of alternating grounds, applied overall lining fabrics that block access to the original canvas, and utilized overall lining adhesives that compromise the analysis of ground and canvas materials. In removing our ability to recognize the underlying "compositions" in Hofmann's ground layers, we lose material evidence of technique that could enhance our understanding of the artist's physical and metaphysical processes. Abstract Expressionist materials play a pivotal role in modern art history, and conservators should be wary of treatments that limit access to or adulterate information related to era-specific product and process. Two years before Hofmann's death, conservator Caroline Keck told an audience of future conservators that "in preserving all art we have two problems: the continuance of its objective, material, structure and the preservation of its integrity of context" (Keck 1964, 1). When approaching the conservation of works by Hofmann and his Abstract Expressionist colleagues, conservators must prioritize both the physical and the art historical impact of the artists' materials and make decisions that carefully balance the preservation of each object with the preservation of that object's role in future scholarship.

Overlooked Areas of Modernist Scholarship

Hofmann recognized that innovative art is built upon a modernist continuum. Through his schools in Europe and the United States, Hofmann distilled the advances of preceding generations of modern painters while supporting the communal dialogue, education, and experimentation that encouraged new forms of modern art. "We are connected with our own age if we recognize ourselves in relation to outside events," Hofmann told his students, "and we have grasped its spirit when we influence the future" (1948, 60). When mural projects in the late 1950s sparked Hofmann's signature painting style and his departure from teaching, the bold color and composition of his subsequent, signature works provided a physical link between nineteenth- and twentieth-century modernism, a continuum recognized by artist Frank Stella, who dubbed Hofmann "the artist of the century": "[We] see Hofmann reach back to the painting of the past with the act of painting, watch as Hofmann pushes his painting to engage with all the painting of his time, and finally, recognize how, by example, Hofmann thrusts painting into the future" (1999, 16).

Hofmann is ultimately an exemplar of modernist spirit—the drive to preserve past lessons while supporting future innovation. He recognized that modernist ideology does not escape eventual historicism. "Art is for me the glorification of the human spirit,"

Hofmann noted, "and as such it is the cultural documentation of the time in which it is produced" (Feinstein 1950). In a telling gesture of his belief in the value of modernist community, when Hofmann agreed to donate forty-seven of his paintings and a quarter of a million dollars to help establish an art museum at the University of California, Berkeley, he included a unique proviso that allowed the university to sell any work from his donated collection in order to purchase art "considered to be avant-garde at the time of such purchase" (BAM/PFA records 1966–present, "Agreement," article 8d).[147] In this manner, Hofmann ensured his ability to bring avant-garde art to the public long after his own era was no longer considered avant-garde. While helping to choose the works for donation, gallerist Samuel Kootz wrote to university faculty—and former Hofmann student—Erle Loran and insisted that the collection include the very latest paintings created by Hofmann, but perhaps the artist's valuation of community over ego is a better example of what Kootz called Hofmann's "brilliant present and promising future" (BAM/PFA records 1966–present, letter, June 8, 1965). Hofmann's last lesson is to restore our understanding of the continuum inherent in modern art—in the community, in the individual artist, in his work, and in his materials. The physical evidence of Hofmann and his modernist colleagues is best appreciated and preserved through our recognition of this continuum.

The legacy of Hofmann and his colleagues would be well served by the continued study of their materials and practices. The extensive analytical data catalogue developed for this study offers the largest and most comprehensive technical study to date on a single Abstract Expressionist artist. As such, it can be mined for more detailed examination of Hofmann's individual materials or compiled with other technical studies to answer questions regarding changing twentieth-century paint formulations—the post–World War II shift from long-oil to short-oil alkyd paint media or the appearance of substitute materials in response to war-era shortage or surplus, for example[148]—or contribute to the long-term study of aging behaviors in modern materials. The data contained in this study can also be used to assess other patterns of materials usage within the mid-twentieth-century New York arts community, particularly patterns that address larger questions about the making and meaning of modern art. One area that has thus far been neglected in technical scholarship is the role of color in the Abstract Expressionist palette.

The innovative use of color is a formative thread throughout modern art. Each of the modernist periods synthesized in Hofmann's late-career paintings is accompanied by the artists' embrace—often with concomitant public and critical rejection—of newly developed paint colors. Hofmann's early art lessons, for example, focused on the style of the Impressionist painters, whose lavish use of new blue and purple pigment formulations caused critics to question whether painting *en plein air* had damaged the artists' eyesight. More innovative, bold paint colors became available during Hofmann's years in Paris, as the "restitution" of symbolic color—lauded by Matisse in 1945 (Matisse 1995b) in the work of avant-garde Fauvist painters and members of Der Blaue Reiter—was greeted with shock and dismay by critics. At the same time, color gained a primary creative function with the emergence of such movements as Orphism and Synchromism. "From its once-secondary position, 'color,'" Guillaume Apollinaire wrote at the time, "is no longer used for just coloring…color is now itself the form.…Color no longer depends on the three dimensions, for it is color that creates them" (Hughes 2012, 75). Color maintained its creative and visual force in the mid-twentieth century in the work of Hofmann and his colleagues. The ability of their modernist palette to shock continues unabated. Hofmann's use of color has been compared to a bomb blast (Stella 1999), while critics from T. J. Clark to Tom Wolfe join Greenberg in exploring the Abstract

Expressionists' potentially purposeful use of "vulgar" color.[149] Yet the primacy of color is rarely discussed in technical scholarship on Abstract Expressionism.

The surge in mid-twentieth-century innovation in paint binding media has shifted the focus of art history and conservation scholarship. Technical studies of early modernist movements acknowledge the formative role played by new color formulations, but the industrial and synthetic paint binders that captured popular and critical attention in the mid-twentieth century have all but pushed aside the artists' own emphasis on color. "The color itself is a plastic medium," Hofmann proclaimed. "Color is a complete medium in itself" (Hofmann 1938–39). While new binding media inarguably played a formative role in the work of some Abstract Expressionist artists, it is now clear that many contemporaneous painters relied heavily on traditional materials while creating their work. Conservation and technical analysis that is focused solely on new paint binders overlooks the parallel advances in color formulation and modernist color that link Hofmann and his colleagues to the rest of art history. By defining modern art history as the appearance of industrial binding media, we break the connections between contemporary art and its influences and replace pivotal movements in modern art history with the myth of the spontaneous genius-artist whose potential contributions fade with his transient fame. The growing compendium of modern art technical studies contains a wealth of information that can be used to look for trends, gaps, and patterns in the use of color by mid-twentieth-century modern painters. It would be particularly interesting to see the patterns of color use that emerge among the Abstract Expressionist painters and track the evolution of modernist color through Hofmann and his Abstract Expressionist colleagues. Rebuilding the Abstract Expressionist link in modernist color would be a step toward reestablishing the modernist community Hofmann embodied.

Appendix I

Tables

TABLE 1 Hofmann Paintings and Palettes Selected for the Study Group

Year	Title	Acc. no.	Size
1947	*Delight*	MoMA 2.1956	127 × 101.6 cm (50 × 40 in.)
	Ecstasy	BAM 1963.2	172.7 × 152.4 cm (68 × 60 in.)
	The Third Hand	BAM 1966.48	152.7 × 101.6 cm (60⅛ × 40 in.)
1953	*Le Gilotin*	BAM 1965.15	147.3 × 121.9 cm (58 × 48 in.)
1954	*Scintillating Space*	BAM 1966.47	213.7 × 122.9 cm (84⅛ × 48⅜ in.)
1955	*Exuberance*	AKAG 1955:8	127 × 101.6 cm (50 × 40 in.)
1957	*Sommernachtstraum*	AKAG 1958:4	132.1 × 152.4 cm (52 × 60 in.)
1958	*Equinox*	BAM 1965.12	183.2 × 153 cm (72⅛ × 60¼ in.)
	Morning Mist	BAM 1966.45	140 × 102.6 cm (55⅛ × 40⅜ in.)
1959	*Above Deep Waters*	BAM 1965.13	214 × 132.1 cm (84¼ × 52 in.)
	Indian Summer	BAM 1965.11	152.7 × 183.4 cm (60⅛ × 72¼ in.)
	Ruby Gold	MAG 60.37	140.7 × 102.9 cm (55⅜ × 40½ in.)
	The Vanquished	BAM 1966.49	91.8 × 122.2 cm (36⅛ × 48⅛ in.)
1960	*Bald Eagle*	BAM 1964.3	153 × 132.7 cm (60¼ × 52¼ in.)
	In the Wake of the Hurricane	BAM 1965.6	183.5 × 152.4 cm (72¼ × 60 in.)
1961	*Combinable Wall I and II*	BAM 1963.10	214.6 × 285.8 cm (84½ × 112½ in.)
	Tormented Bull	BAM 1963.6	152.7 × 214 cm (60⅛ × 84¼ in.)
1962	*Heraldic Call*	BAM 1965.17	153 × 122.9 cm (60¼ × 48⅜ in.)
	Magnum Opus	BAM 1963.7	213.7 × 198.4 cm (84⅛ × 78⅛ in.)
	Memoria in Aeternum	MoMA 399.1963	213.3 × 183.2 cm (84 × 72⅛ in.)
1963	*Polyhymnia*	BAM 1964.1	183.2 × 153 cm (72⅛ × 60¼ in.)
1964	*The Clash*	BAM 1965.8	132.4 × 153 cm (52⅛ × 60¼ in.)
	Imperium in Imperio	BAM 1966.43	213.7 × 132.1 cm (84⅛ × 52 in.)
	And Out of the Caves the Night Threw a Handful of Pale Tumbling Pigeons into the Light	BAM 1965.4	213.7 × 153 cm (84⅛ × 60¼ in.)
	Silent Night	BAM 1965.5	213.3 × 198.8 cm (84 × 78¼ in.)
1965	*Struwel Peter*	BAM 1966.5	183.2 × 153 cm (72⅛ × 60¼ in.)
1966	Palette on plywood	Trust M536-12	19.1 × 19.1 cm (7½ × 7½ in.)
	Palette on board	Trust M593-12	12.1 × 22.9 cm (4¾ × 9 in.)
	Palette on board	Trust M537-10	104.1 × 20.3 cm (41 × 8 in.)
	Palette on board	Trust M536-53	16.5 × 20.3 cm (6½ × 8 in.)
	Palette on glass	Trust [no #]	61 × 61 cm (24 × 24 in.)
	Palette on paint can lid	Trust M536-49	15.2 cm (6 in.) diameter
	Palette on board	Trust M536-45	15.2 × 29.9 cm (6 × 11¾ in.)
	Palette on board	Trust M536-03	10.2 × 27.9 cm (4 × 11 in.)

AKAG: Albright-Knox Art Gallery (Buffalo, NY)
BAM: University of California, Berkeley Art Museum and Pacific Film Archive
MAG: Memorial Art Gallery, University of Rochester (Rochester, NY)
MoMA: Museum of Modern Art (New York)
Trust: Renate, Hans and Maria Hofmann Trust (New York)

TABLE 2 Condition Issues and Evidence of the Artist's Technique Observed in the Study Group Paintings

Year	Title	Acc. no.	Distortions	Localized distortions	Cracking/cleaving paint	Pinholes/tape marks	Surface wrinkling	Crystalline exudates, black paints	Crystalline exudates, other colors	Soft paint/liquid exudates, magenta	Soft paint/liquid exudates, other colors	Underbound/desiccated paints
1947	*Delight*	MoMA 2.1956			X							X
	Ecstasy	BAM 1963.2	X		X							X
	The Third Hand	BAM 1966.48			X							X
1953	*Le Gilotin*	BAM 1965.15	X		X					X		
1954	*Scintillating Space*	BAM 1966.47	X		X							
1955	*Exuberance*	AKAG 1955:8	X		X							
1957	*Sommernachtstraum*	AKAG 1958:4	X		X				X			
1958	*Equinox*	BAM 1965.12	X		X	X				X		
	Morning Mist	BAM 1966.45		X	X	X						X
1959	*Above Deep Waters*	BAM 1965.13	X		X				X			X
	Indian Summer	BAM 1965.11		X	X				X	X		X
	Ruby Gold	MAG 60.37		X	X					X		X
	The Vanquished	BAM 1966.49	X		X		X					X
1960	*Bald Eagle*	BAM 1964.3	X		X		X	X	X	X		X
	In the Wake of the Hurricane	BAM 1965.6	X		X					X		X
1961	*Combinable Wall I and II*	BAM 1963.10		X	X	X			X		X	
	Tormented Bull	BAM 1963.6			X	X						
1962	*Heraldic Call*	BAM 1965.17			X			X	X			
	Magnum Opus	BAM 1963.7			X	X		X	X	X		
	Memoria in Aeternum	MoMA 399.1963			X	X						
1963	*Polyhymnia*	BAM 1964.1			X	X		X		X	X	X
1964	*The Clash*	BAM 1965.8	X		X		X		X	X		X
	Imperium in Imperio	BAM 1966.43	X		X	X			X			X
	And Out of the Caves . . .	BAM 1965.4	X		X	X			X	X		
	Silent Night	BAM 1965.5	X		X	X						
1965	*Struwel Peter*	BAM 1966.5			X				X	X	X	X

Note: Information regarding the condition and conservation of the study group works was compiled from the conservation records of the respective collections and from in situ assessment of the paintings during study group selection and sampling.

AKAG: Albright-Knox Art Gallery (Buffalo, NY)

BAM: University of California, Berkeley Art Museum and Pacific Film Archive

MAG: Memorial Art Gallery, University of Rochester (Rochester, NY)

MoMA: Museum of Modern Art (New York)

TABLE 3 Conservation Materials Mentioned in the Treatment Documentation

			Lining / consolidation					Varnishing / inpainting						Resaturation	
Year	Title	Acc. no.	Wax-resin	PVA	pEVA[a]	pnBMA	Other[b]	pnBMA	piBMA	p(EMA-MA)	PVA	Acrylic paints	Aqueous media[c]	Acrylic resins[d]	Polycyclohexanone
1947	*Delight*	MoMA 2.1956		X				X							
	Ecstasy	BAM 1963.2	X		X		X		X						
	The Third Hand	BAM 1966.48	X	X	X				X				X		
1953	*Le Gilotin*	BAM 1965.15	X	X	X			X	X	X					
1954	*Scintillating Space*	BAM 1966.47	X		X		X	X	X	X		X			
1955	*Exuberance*	AKAG 1955:8	X		X						X		X		
1957	*Sommernachtstraum*	AKAG 1958:4			X								X		
1958	*Equinox*	BAM 1965.12	X	X	X		X	X	X	X					
	Morning Mist	BAM 1966.45	X		X		X						X		
1959	*Above Deep Waters*	BAM 1965.13		X			X	X	X						
	Indian Summer	BAM 1965.11													
	Ruby Gold	MAG 60.37		X							X			X	
	The Vanquished	BAM 1966.49			X	X	X	X	X			X	X		
1960	*Bald Eagle*	BAM 1964.3	X		X			X	X						X
	In the Wake of the Hurricane	BAM 1965.6			X			X	X						
1961	*Combinable Wall I and II*	BAM 1963.10		X											
	Tormented Bull	BAM 1963.6	X		X		X	X	X				X		
1962	*Heraldic Call*	BAM 1965.17													
	Magnum Opus	BAM 1963.7						X							
	Memoria in Aeternum	MoMA 399.1963			X							X			
1963	*Polyhymnia*	BAM 1964.1			X										
1964	*The Clash*	BAM 1965.8		X	X		X	X	X	X					
	Imperium in Imperio	BAM 1966.43			X	X									
	And Out of the Caves . . .	BAM 1965.4	X					X	X						
	Silent Night	BAM 1965.5	X						X				X		
1965	*Struwel Peter*	BAM 1966.5	X	X	X		X						X		

Note: Information provided by archived conservation treatment reports is not complete; additional materials may have been used prior to or during treatments. Volatile conservation materials (solvents, etc.) are not listed.

AKAG: Albright-Knox Art Gallery (Buffalo, NY)
BAM: University of California, Berkeley Art Museum and Pacific Film Archive
MAG: Memorial Art Gallery, University of Rochester (Rochester, NY)
MoMA: Museum of Modern Art (New York)
Wax-resin: A mixture of natural and synthetic waxes and synthetic resins
PVA: Poly(vinyl acetate), noted as AYAA and AYAC formulations in consolidation, AYAB for inpainting, AYAA and AYAC in resaturation
pEVA: Poly(ethylene vinyl acetate), noted as BEVA 371 and BEVA D-8 in consolidation
p*n*BMA: Poly(n-butyl methacrylate), noted as Bedacryl 122X in consolidation, Lucite 44 and Solvar in varnishing, Magna Plastic Colors in inpainting
p*i*BMA: Poly(isobutyl methacrylate), noted as Paraloid B-67 in varnishing and inpainting
p(EMA-MA): Poly(ethyl methacrylate methyl acrylate), noted as Paraloid B-72 in varnishing and inpainting
Acrylic paints: Noted include Golden MSA Conservation Paints and Lefranc and Bourgeois Restoration Colors (acrylic/ketonic mixture)
Polycyclohexanone: A ketonic resin noted as Winton Retouch Varnish in resaturation

[a] As a component of BEVA 371 and BEVA D-8 adhesives, also contains polycyclohexanone resin and wax.
[b] Local fill materials noted include adhesive putty, latex spackle, and Polyfix; old glue lining noted on *Equinox*.
[c] Including gouache, watercolors, and pastels.
[d] Noted as "various synthetic resins."

TABLE 4 Complementary Analytical Techniques Used in the Identification of Hofmann's Materials

	Chemical characterization		Materials identified		Spatial resolution			Complementary information				
	Elemental	Molecular	Pigments	Binding media	Pigment particle	Individual paint layer	Multiple paint layers	Optical microscopy	SEM-EDS	XRD	Py-GC-MS[a]	FTIR
Visual analysis												
Optical microscopy	–	–	X	X		X	X		X	X	X	X
Scanning electron microscopy (SEM)	–	–	X	X	X	X	X	X		X		
Inorganic analysis												
Energy dispersive spectroscopy (EDS)	X		X		X	X		X		X		
X-ray diffraction (XRD)		X	X		X			X	X			
Organic analysis												
Gas chromatography–mass spectrometry (GC-MS)[b]		X	X	X		X	X	X				X
Fourier transform infrared spectroscopy (FTIR)[b]		X	X	X		X	X	X			X	

[a] Pyrolysis is commonly coupled with GC-MS for the analysis of large-molecule modern paint binders. GC-MS can also be used to identify some organic pigments but cannot identify inorganic pigments.

[b] Analysis of individual paint layers by GC-MS or FTIR requires separation of the paint layers prior to analysis.

TABLE 5 Number of Analysis Locations for Each Study Group Painting and Palette

Year	Title	Acc. no.	Sampling[a]	SEM-EDS[b]	Py-GC-MS	FTIR[c]	XRD
1947	*Delight*	MoMA 2.1956	9	19	4	6	—
	Ecstasy	BAM 1963.2	9	18	4	3	—
	The Third Hand	BAM 1966.48	10	16	5	10	—
1953	*Le Gilotin*	BAM 1965.15	11	39	—	19	—
1954	*Scintillating Space*	BAM 1966.47	9	16	2	9	—
1955	*Exuberance*	AKAG 1955:8	11	16	7	10	—
1957	*Sommernachtstraum*	AKAG 1958:4	10	14	5	10	—
1958	*Equinox*	BAM 1965.12	11	19	2	12	2
	Morning Mist	BAM 1966.45	10	16	—	13	—
1959	*Above Deep Waters*	BAM 1965.13	12	23	6	15	2
	Indian Summer	BAM 1965.11	9	14	2	10	3
	Ruby Gold	MAG 60.37	13	25	8	11	2
	The Vanquished	BAM 1966.49	10	35	3	13	—
1960	*Bald Eagle*	BAM 1964.3	13	21	8	11	—
	In the Wake of the Hurricane	BAM 1965.6	9	30	4	10	2
1961	*Combinable Wall I and II*	BAM 1963.10	13	24	1	13	4
	Tormented Bull	BAM 1963.6	4	6	4	3	—
1962	*Heraldic Call*	BAM 1965.17	4	5	4	—	—
	Magnum Opus	BAM 1963.7	6	9	4	6	—
	Memoria in Aeternum	MoMA 399.1963	11	25	2	9	—
1963	*Polyhymnia*	BAM 1964.1	6	14	3	5	1
1964	*The Clash*	BAM 1965.8	9	14	4	10	—
	Imperium in Imperio	BAM 1966.43	8	16	—	7	—
	And Out of the Caves . . .	BAM 1965.4	5	13	2	6	1
	Silent Night	BAM 1965.5	14	24	6	17	—
1965	*Struwel Peter*	BAM 1966.5	11	18	9	10	—
1966	Palette on plywood	Trust M536-12	1	1	1	1	—
	Palette on board	Trust M593-12	1	5	—	1	—
	Palette on board	Trust M537-10	8	16	—	9	1
	Palette on board	Trust M536-53	5	9	—	6	4
	Palette on glass	Trust [no #]	5	8	3	6	—
	Palette on paint can lid	Trust M536-49	3	6	2	4	—
	Palette on board	Trust M536-45	3	6	1	4	—
	Palette on board	Trust M536-03	4	6	1	5	—

AKAG: Albright-Knox Art Gallery (Buffalo, NY)
BAM: University of California, Berkeley Art Museum and Pacific Film Archive
MAG: Memorial Art Gallery, University of Rochester (Rochester, NY)
MoMA: Museum of Modern Art (New York)
Trust: Renate, Hans and Maria Hofmann Trust (New York)

SEM-EDS: Scanning electron microscopy–energy dispersive spectroscopy
Py-GC-MS: Pyrolysis–gas chromatography–mass spectrometry
FTIR: Fourier transform infrared spectroscopy
XRD: X-ray diffraction

[a] Cross sections are counted here as a single sample. Figure does not include fiber samples.
[b] Three points within each analysis location were examined by EDS.
[c] FTIR analysis did not differentiate between cross-section layers; large cross sections were analyzed in sections.

TABLE 6 Pigments Identified by Energy Dispersive Spectroscopy

Year	Title	Acc. no.	Zinc white	Titanium white	Cadmium yellow	Yellow ocher	Zinc yellow	Cadmium orange	Cadmium red	Red ocher	Cadmium green	Phthalocyanine green	Viridian/chrome green	Iron oxide green	Cobalt green	Ultramarine blue	Phthalocyanine blue	Cobalt blue	Cerulean blue	Prussian blue	Ultramarine violet	Cobalt violet	Ivory black	Carbon black[a]
1947	*Delight*	MoMA 2.1956	X		X	X			X					X		X								X
	Ecstasy	BAM 1963.2	X		X		X		X			X				X	X						X	
	The Third Hand	BAM 1966.48	X		X				X			X				X								
1953	*Le Gilotin*	BAM 1965.15	X		X				X		X	X				X	X					X	X	
1953	*Scintillating Space*	BAM 1966.47	X	X	X			X	X		X	X				X								
1953	*Exuberance*	AKAG 1955:8	X	X	X				X		X					X								
1953	*Sommernachtstraum*	AKAG 1958:4	X	X	X	X		X	X	X	X					X				X				
1958	*Equinox*	BAM 1965.12	X		X		X	X	X		X	X				X			X			X		
	Morning Mist	BAM 1966.45	X		X	X		X	X		X					X	X				X			
1959	*Above Deep Waters*	BAM 1965.13	X	X	X			X	X		X	X				X							X	
	Indian Summer	BAM 1965.11	X		X			X	X		X					X								
	Ruby Gold	MAG 60.37	X	X	X			X	X		X			X		X						X		X
	The Vanquished	BAM 1966.49	X		X			X	X		X	X				X						X		X
1960	*Bald Eagle*	BAM 1964.3	X	X	X			X	X		X	X				X		X				X	X	X
	In the Wake of the Hurricane	BAM 1965.6	X	X	X			X	X		X					X						X		
1961	*Combinable Wall I and II*	BAM 1963.10	X	X	X	X		X	X		X	X				X							X	
	Tormented Bull	BAM 1963.6	X	X																				X
1962	*Heraldic Call*[b]	BAM 1965.17		X					X														X	
	Magnum Opus	BAM 1963.7	X	X	X				X							X		X						
	Memoria in Aeternum	MoMA 399.1963	X	X	X			X	X		X				X	X								
1963	*Polyhymnia*[b]	BAM 1964.1	X	X					X		X					X							X	
1964	*The Clash*	BAM 1965.8	X	X	X			X	X			X				X						X		
	Imperium in Imperio[b]	BAM 1966.43	X		X			X	X		X	X				X		X						
	And Out of the Caves . . .	BAM 1965.4	X	X		X			X		X	X		X								X		
	Silent Night	BAM 1965.5	X	X	X	X		X			X	X	X		X	X		X						
1965	Struwel Peter	BAM 1966.5	X	X	X			X	X		X					X						X		
1966	Palette on plywood[b]	Trust M536-12	X																					
	Palette on board[b]	Trust M593-12	X									X												
	Palette on board[b]	Trust M537-10	X		X			X	X		X	X				X								
	Palette on board[b]	Trust M536-53	X					X	X															
	Palette on glass[b]	Trust [no #]	X		X	X					X					X	X							
	Palette on paint can lid[b]	Trust M536-49	X	X												X								
	Palette on board[b]	Trust M536-45	X	X												X				X				
	Palette on board[b]	Trust M536-03		X								X												

AKAG: Albright-Knox Art Gallery (Buffalo, NY)
BAM: University of California, Berkeley Art Museum and Pacific Film Archive
MAG: Memorial Art Gallery, University of Rochester (Rochester, NY)
MoMA: Museum of Modern Art (New York)
Trust: Renate, Hans and Maria Hofmann Trust (New York)

[a] Carbon black is not detectable by EDS and may be present along with other black pigments.
[b] Not all paint materials were sampled from this study group item.

TABLE 7 Pigments Identified by Fourier Transform Infrared Spectroscopy

			Phthalocyanine blue					Phthalocyanine green		Alizarin		Rhodamine
Year	Title	Acc. no.	PB15	PB15:1	PB15:2	PB15:4	PB15:6	PG7	Likely[a]	PR83	Likely[b]	PR81
1947	*Delight*	MoMA 2.1956									X	
	Ecstasy	BAM 1963.2									X	
	The Third Hand	BAM 1966.48								X		
1953	*Le Gilotin*	BAM 1965.15	X	X				X	X	X		
1954	*Scintillating Space*	BAM 1966.47						X				
1955	*Exuberance*	AKAG 1955:8										
1957	*Sommernachtstraum*	AKAG 1958:4							X			
1958	*Equinox*	BAM 1965.12						X				
	Morning Mist	BAM 1966.45	X	X					X	X		
1959	*Above Deep Waters*	BAM 1965.13										
	Indian Summer	BAM 1965.11						X				
	Ruby Gold	MAG 60.37										
	The Vanquished	BAM 1966.49										
1960	*Bald Eagle*	BAM 1964.3										
	In the Wake of the Hurricane	BAM 1965.6										
1961	*Combinable Wall I and II*	BAM 1963.10							X			
	Tormented Bull	BAM 1963.6										
1962	*Heraldic Call*[c]	BAM 1965.17										
	Magnum Opus	BAM 1963.7									X	
	Memoria in Aeternum	MoMA 399.1963										
1963	*Polyhymnia*[c]	BAM 1964.1									X	
1964	*The Clash*	BAM 1965.8								X		
	Imperium in Imperio[c]	BAM 1966.43							X			
	And Out of the Caves . . .	BAM 1965.4										
	Silent Night	BAM 1965.5						X				
1965	*Struwel Peter*	BAM 1966.5										
1966	Palette on plywood[c]	Trust M536-12										
	Palette on board[c]	Trust M593-12	X	X	X	X	X					
	Palette on board[c]	Trust M537-10						X				
	Palette on board[c]	Trust M536-53										X
	Palette on glass[c]	Trust [no #]	X	X			X					
	Palette on paint can lid[c]	Trust M536-49										X
	Palette on board[c]	Trust M536-45										
	Palette on board[c]	Trust M536-03						X	X			X

AKAG: Albright-Knox Art Gallery (Buffalo, NY)
BAM: University of California, Berkeley Art Museum and Pacific Film Archive
MAG: Memorial Art Gallery, University of Rochester (Rochester, NY)
MoMA: Museum of Modern Art (New York)
Trust: Renate, Hans and Maria Hofmann Trust (New York)

[a] Samples not directly identified as the pigment but matching FTIR spectra for other samples containing the pigment.
[b] Matched to other samples identified as PR83 or presenting SEM-EDS results consistent with an alumina substrate.
[c] Not all paint materials were sampled from this study group item.

TABLE 8 Summary of Conservation Materials Identified in Study Group Samples

Year	Title	Acc. no.	Sample no.	Result	Method(s)
1947	*Ecstasy*	BAM 1963.2	Esc01	Wax	Py-GC-MS
			Esc02	Acrylic, wax	Py-GC-MS
			Esc03	Acrylic	Py-GC-MS
			Esc06	Acrylic varnish	ATR-FTIR
			Esc07	Wax	Py-GC-MS
	The Third Hand	BAM 1966.48	C063	Acrylics	Py-GC-MS
			C064	Acrylic	Py-GC-MS
			C071	Wax	Py-GC-MS, ATR-FTIR
1953	*Le Gilotin*	BAM 1965.15	C016	Acrylic varnish	ATR-FTIR
1954	*Scintillating Space*	BAM 1966.47	C072	Wax	ATR-FTIR
			C074	Acrylic varnish	ATR-FTIR
			C076	Acrylic varnish	ATR-FTIR
			C077	Acrylic varnish	ATR-FTIR
			C078	Acrylic varnish	ATR-FTIR
1958	*Equinox*	BAM 1965.12	C050	Acrylic varnish	ATR-FTIR
			C052	Acrylic varnish	ATR-FTIR
			C053	Acrylic varnish	ATR-FTIR
			C057	Acrylic media	ATR-FTIR
1959	*Above Deep Waters*	BAM 1965.13	C110	Wax	ATR-FTIR
			C111	Wax	ATR-FTIR
			C114	Wax	ATR-FTIR
			C117	Wax	ATR-FTIR
1960	*Bald Eagle*	BAM 1964.3	C027	Acrylic varnish	ATR-FTIR
			C030	Acrylic varnish	Py-GC-MS, ATR-FTIR
			C033	Acrylic	Py-GC-MS
			C036	Acrylic varnish	ATR-FTIR
	In the Wake of the Hurricane	BAM 1965.6	C175	Acrylic varnish	ATR-FTIR
			C176	Acrylic	Py-GC-MS
			C178	Acrylic	Py-GC-MS
			C180	Wax	Py-GC-MS
1961	*Combinable Wall I and II*	BAM 1963.10	C083	Acrylic varnish	ATR-FTIR
			C084	Acrylic varnish	ATR-FTIR
			C089	Acrylic varnish	ATR-FTIR
	Tormented Bull	BAM 1963.6	C061	Acrylics	Py-GC-MS
1962	*Magnum Opus*	BAM 1963.7	C154	Acrylic varnish	ATR-FTIR

Year	Title	Acc. no.	Sample no.	Result	Method(s)
1964	*The Clash*	BAM 1965.8	C004	Acrylic, wax	Py-GC-MS, ATR-FTIR
			C005	Acrylic varnish	ATR-FTIR
			C006	Wax	ATR-FTIR
			C011	Acrylic	Py-GC-MS
	And Out of the Caves . . .	BAM 1965.4	C170	Acrylic varnish	ATR-FTIR
			C172	Acrylic	Py-GC-MS
	Silent Night	BAM 1965.5	C104	Acrylic, wax	Py-GC-MS
			C105	Acrylic varnish	ATR-FTIR
			C106	Acrylic varnish	ATR-FTIR
			C107	Acrylic varnish	ATR-FTIR
			C108	Acrylic varnish	ATR-FTIR
			C109	Acrylic varnish	Py-GC-MS, ATR-FTIR

BAM: University of California, Berkeley Art Museum and Pacific Film Archive
Py-GC-MS: Pyrolysis–Gas Chromatography–Mass Spectrometry
ATR-FTIR: Attenuated Total Reflectance–Fourier Transform Infrared Spectroscopy

TABLE 9 Pigment Polymorphs Identified by X-Ray Diffraction

Year	Title	Acc. no.	Sample no.	Description	Cadmium sulfide selenide[a]	Cadmium selenide sulfide[b]	Cadmium sulfide, CdS[c]	Barite, $BaSO_4$	Calcium sulfate, $CaSO_4$	Zincite, $(Zn, Mn)O$	Anatase, TiO_2	Lazurite[d]	Cobalt phosphate, $Co_3(PO_4)_2$[e]
1958	*Equinox*	BAM 1965.12	C047	Dark red	X			X					
			C056	Purple				X					X
1959	*Above Deep Waters*	BAM 1965.13	C110	Primary red	X			X					
			C111	Secondary red	X			X					
	Indian Summer	BAM 1965.11	C128	Main orange		X	X						
			C129	Purple									X
			C131	Bright red	X			X					
	Ruby Gold	MAG 60.37	R03	Orange[f]	X			X					
			R11	Orange[f]	X			X					
1960	*In the Wake of the Hurricane*	BAM 1965.6	C174	Purple									X
			C175	Unknown blue								X	
1961	*Combinable Wall I and II*	BAM 1963.10	C087	Main red	X			X	X	X	X		
			C088	Secondary red	X			X		X	X		
			C089	Pink	X			X		X	X		
			C092	Main orange	X			X		X	X		
1963	*Polyhymnia*	BAM 1964.1	C133	Yellow			X						
1964	*And Out of the Caves . . .*	BAM 1965.4	C173	Purple									X
1966	Palette on board	Trust M537-10	S10	Red	X			X		X	X		
	Palette on board	Trust M536-03	S11	Red	X			X					
	Palette on board	Trust M536-53	S13	Dark red[g]				X	X				
			S13	Dark orange	X			X	X				
			S14	Orange		X	X						

BAM: University of California, Berkeley Art Museum and Pacific Film Archive
MAG: Memorial Art Gallery, University of Rochester (Rochester, NY)
Trust: Renate, Hans and Maria Hofmann Trust (New York)

[a] Cadmium sulfide selenide polymorph identified in International Center for Diffraction Data (ICDD) libraries as $CdS_{0.54}Se_{0.46}$.
[b] Cadmium selenide sulfide polymorph identified in ICDD libraries as $CdS_{0.42}Se_{0.58}$.
[c] Trace levels of manganese in the cadmium sulfide pigment in *Polyhymnia* produced a cadmium manganese sulfide polymorph identified in ICDD libraries as $(Cd_{0.65}Mn_{0.35})S$.
[d] Lazurite polymorph identified in ICDD libraries as $Na_{6.56}Ca_{1.52}(AlSiO_4)_6(SO_4)_{0.84}S_{1.544}$.
[e] Cobalt phosphate appears in conjunction with dittmarite; in some cases, dittmarite is the only material found in samples identified as cobalt phosphate through other analyses.
[f] Samples taken from an area of localized distortion and from an area at the edge of the composition that had no visible ground layer.
[g] Identified by FTIR as *Colour Index* Pigment Red 81.

Appendix II

Sample Preparation and Instrumentation Setup

Cross-Section Preparation

Samples were positioned on precast epoxy half-tablets and adhered using a small droplet of cyanoacrylate adhesive. The sample and half-tablet were then transferred into a silicone rubber mold for embedding. The epoxy resin, Tra-bond 2113 two-part system, was mixed and poured over the adhered sample and half-tablet. The liquid resin was held at room temperature for 1 hour, then cured at approximately 45°C for 3 hours. Once demolded, the tablets were trimmed with a mill to expose the sample. Aluminum oxide abrasive was used to polish the face and to edge the surfaces with an aliphatic hydrocarbon used as a lubricant. (Some samples were dry-polished later for FTIR analysis.) Of the 284 paint samples, 83 presented multilayered stratigraphy and were selected for mounting and analysis as cross sections.

Instrumentation Setup

Optical Microscopy

Samples were viewed on Leica models DMLM and DMRX research microscopes. Both share some components, such as objectives and filters. The range of magnification for the reflected light techniques is 50x to 500x, measured at the eyepieces. Objectives are 5x, 10x, 20x, 40x, and 50x. Through the use of a prism system, brightfield and darkfield illumination conditions can be alternated. The tungsten halogen lamp is the primary source of reflected light, and it is corrected for daylight color temperature using a dichroic mirror. The fluorescence illuminator is either a 100 W mercury or a 150 W xenon lamp. The filters span the range from ultraviolet to the violet. The characteristics of the Leitz (Leica) filter sets are provided in table A2.1.

A purpose-built digital camera is used for image acquisition (camera chip dimensions, as well as pixel count, determine ultimate magnification of image). The camera system is a Nikon DMX 1200 24-bit color system. The camera uses the Sony ICX085AK color CCD and Nikon's proprietary Inter Pixel Stepping (IPS) high-density imaging technology. Images are normally captured as 24-bit 1.4 MP tif images, approximately 3.8 MP each.

All 284 paint material samples and 28 fiber samples were examined and photographed under both normal and ultraviolet illumination sources. Multilayered paint samples were embedded as cross sections prior to photography. The remaining paint samples were photographed in their storage vials and subsequently analyzed in their loose form.

Pyrolysis–Gas Chromatography–Mass Spectrometry

Samples were analyzed with a Varian Saturn 2000 GC/MS equipped with a CDS Pyroprobe 2000. Each sample was derivatized with 2 μL of tetramethylammonium hydroxide (TMAH), which was put onto the sample in a quartz boat. The boat was placed into the coiled platinum probe of a CDS Pyroprobe 2000 filament pyrolysis unit, and the probe was then placed into a helium-purged CDS 1500 Valved Interface attached to the Varian GC. The interface was held at a constant 310°C and purged with helium for 10 seconds before the valve to the GC column was opened. The sample was then heated with the pyroprobe to a temperature of approximately 600°C for 10 seconds. The pyrolysis products were transferred directly to a capillary column (ZB-5ms; 30 m x 0.25 mm i.d.; 0.25 μm film thickness; He flow of 1.2 mL/min; splitless) in a Varian 3800 gas chromatograph (GC) equipped with electronic flow control. The GC oven was programmed with an initial temperature of 30°C, which was held for 5 minutes. The temperature was increased at

TABLE A2.1 Leitz Filters for Fluorescence Microscopy (DMRX and DMLM)

Filter system	Excitation	Excitation filter	Split mirror	Suppression filter	Part no.
A	UV	BP 340–380	RKP 400	LP 430	513 804
D	Violet	BP 355–425	RKP 455	LP 460	513 805

UV: ultraviolet
BP: band pass filter
RKP: reflection short pass filter
LP: long pass filter

a rate of 10°C per minute to 300°C and held for 10 minutes. The Varian 3800 GC was interfaced to a Varian Saturn 2000 ion trap, the transfer line being held at 270°C. Operating conditions for the trap were: trap 150°C, manifold at 80°C; electron multiplier 1500 V; scan range 45–650 amu; scan time 1 second; data analysis Saturn GC/MS Workstation 6.5 software and the NIST 2005 spectral libraries.

GC-MS binding media analysis was performed on 106 discrete paint samples. Samples of Hofmann's materials selected for Py-GC-MS analysis included materials expected to contain modern polymer formulations—ground layer materials reported by others to contain house paints—and paints exhibiting the flow characteristics or surface sheen associated with industrial paints. All white and black paints were analyzed with Py-GC-MS in order to differentiate suspected industrial paints in Hofmann's palette from traditional materials of similar color.

Scanning Electron Microscopy–Energy Dispersive Spectroscopy

Samples were imaged and analyzed with a Hitachi S3700-N scanning electron microscope and a Bruker XFlash energy dispersive spectrometer with Quantax 400 software. The samples were received after they had been prepared for and imaged by polarized light microscopy. They were carbon-coated before analysis. The SEM was operated at 15 kV at variable pressure (40 Pa). The system was calibrated to the setting of 130K counts per second, as improper peak shift was detected at the 275K setting. Elemental maps were generated over 180 seconds real time (with 0%–18% dead time). Analyses were conducted at a working distance between 9.8 and 10.2 mm.

Backscatter electron images were taken of all 284 loose and mounted paint samples, and EDS inorganic materials analysis was performed on each of the 519 discernible paint layers within those samples. EDS analysis was performed on at least three disparate points within each paint layer. EDS analysis was also performed on visible inclusions, such as mordant materials for dye-based pigments.

Fourier Transform Infrared Spectroscopy

Fourier transform infrared spectroscopy in attenuated total reflectance (ATR-FTIR) analysis of loose samples was performed with a Thermo Nicolet 6700 Fourier transform infrared (FTIR) spectrometer with a Golden Gate ATR with diamond crystal, single bounce (45°) sampling accessory and DTGS detector. Spectra were obtained from 64 scans taken at a resolution of 4 cm^{-1}. Samples for ATR-FTIR were placed directly on the diamond crystal of the ATR accessory. For small samples, a piece of aluminum foil was used to back the sapphire anvil to eliminate any sapphire absorption in the infrared spectrum. Micro-FTIR (μFTIR) infrared microscope analysis of embedded cross sections was performed with a Thermo Nicolet 6700 Fourier transform infrared spectrometer with a Continuum microscope, MCT/A detector, and a single bounce diamond crystal μATR objective. Spectra were obtained from 128 scans taken at a resolution of 4 cm^{-1} between 4000 and 625 cm^{-1}. Samples for micro-FTIR were examined directly with the use of a variable aperture so that the area of interest could be chosen. All spectra were ATR corrected and examined using OMNIC v 7.2 and compared to Infrared and Raman Users Group (IRUG) 2000 and local spectral libraries.

A total of 248 paint samples from the study group paintings were analyzed with FTIR. Embedded cross-section samples were repolished to remove the carbon coating from previous SEM-EDS analysis; they were then examined with the use of an FTIR microscope attachment (micro-FTIR). Useful FTIR data were acquired from 76 repolished cross sections. Loose paint samples were examined with an attenuated total reflectance (ATR-FTIR) platform. FTIR analysis of loose sample was performed on excess sample from 167 loose paints, as well as excess sample of the same materials appearing in the 5 cross sections deemed too small for micro-FTIR.

X-Ray Diffraction

Samples were analyzed with Rigaku D/Max Rapid Micro X-Ray Diffractometer and Rigaku AreaMax 2.0 software. Samples for XRD analysis were mounted on glass fiber with Elmer's glue on a sample holder. The copper Kα collimator tube with a 0.8 mm aperture was operated at 50 kV and 40 mA to achieve 2.00 kW. The goniometer was set with an omega angle at 45° while the sample was adjusted on two different axes—a chi fixed at 0° and a phi determined by the position and orientation of the desired analysis site, so that orientation of the crystals with respect to the beam could be randomized as much as possible. The phi axis could remain at a fixed position, oscillate within a degree range, or fully rotate at a speed of 1° per second. XRD patterns were produced of sample areas with diameters of 100–800 μm and with a penetration depth of 20–100 μm, depending on the density and mass of elements as well as on the angle of incidence. After background subtraction, diffraction patterns produced for the samples were qualitatively matched with Jade 8.0 software to reference patterns of known materials in the International Center for Diffraction Data (ICDD) libraries and/or user libraries developed from reference materials.

A total of 22 paint samples from the study group paintings were analyzed with XRD. The XRD sample group includes 11 examples of cadmium paint appearing on a total of five paintings produced between 1958 and 1961, 4 examples of cadmium paint appearing on three of the Estate palettes, and 7 samples of paint whose identification through other analytical methods had produced limited or inconsistent results.

Notes

Introduction

1 T. J. Clark, lecture at the University of California, Berkeley Art Museum, October 15, 1992. This lecture was the starting point for Clark's essay "In Defense of Abstract Expressionism" (1994).

2 Access to the microscopy, SEM-EDS, FTIR, and XRD facilities at the Museum Conservation Institute, Smithsonian Institution, was granted on March 10, 2011, and on August 22, 2013. Access to the Py-GC-MS facilities at the National Gallery of Art was granted on August 4, 2011. Permission to sample selected paintings from the collection of the University of California, Berkeley Art Museum and Pacific Film Archive was granted on April 25, 2011. Permission to sample selected paintings from the collection of the Museum of Modern Art was granted on August 13, 2010. Permission to sample selected paintings from the collection of the Albright-Knox Art Gallery was granted on November 2, 2011. Permission to sample paintings from the University of Rochester's Memorial Art Gallery was granted on October 4, 2011. Access to sample selected palettes from the collection of the Renate, Hans and Maria Hofmann Trust was granted on October 26 and November 7, 2011.

Chapter 1

EPIGRAPHS

Rosenberg 1952, 22–23; Hofmann, 1966, quoted in Jaffe 1971, 37.

3 A comprehensive account of Hofmann's history is found in Dickey 2011, *Color Creates Light: Studies with Hans Hofmann*, in which Hofmann scholar Tina Dickey correlates and documents myriad details of the artist's life and teaching career.

4 Although Hofmann is reported to have alternately embraced and rejected Hildebrand's text, *The Problem of Form* (first translated into English in 1907) was assigned reading for Hofmann's students through the 1930s; it was replaced by Sheldon Cheney's *Expressionism in Art* (1948). The Hildebrand book requirement is cited in Goodman 1990, 78, 97. Cheney's book is listed on a 1940s supply list distributed to Hofmann's students (Kiesler papers).

5 Michailow is also spelled Michaelof (Wolf 1945, 52) and Michaelow (Goodman 1986, 16), and cultural variations on the spelling of Schwarz's first name (Willie, Willi, Willy) appear throughout the literature.

6 NKVM was founded in 1909. The other founding members were Munich art students Marianne von Werefkin, Gabriele Münter, Adolf Erbslöh, and Alexander Kanoldt. NKVM exhibitions were held in 1909, 1910, and 1911. Members of Der Blaue Reiter held a parallel exhibition in 1911 and were not included in the NKVM exhibition. Kandinsky and Jawlensky were also founding members of Der Blaue Reiter. Other founding members included von Werefkin, Münter, Franz Marc, and August Macke.

7 Freudenberg is alternately referred to in the literature as the owner of the department store or as the owner's son.

8 *Gil Blas* critic Louis Vauxcelles is said to have seen a classical sculpture by Albert Marquet surrounded by the new paintings at the 1905 Salon d'Automne and exclaimed, "Donatello parmi les fauves!" (Donatello among the wild beasts!).

9 Matisse operated his school from 1908 to 1911 at 56 rue de Sèvres. The Académie Julian was established by Rodolphe Julian in 1868 and merged with ESAG Penninghen in 1968.

10 The Académie de la Grande Chaumière was founded in 1902 by Martha Stettler and operated open drawing classes at 14 rue de la Grande Chaumière until its sale to the Charpentier family in 1957; the school is now called the Académie de la Charpentier. The Académie Colarossi was founded by sculptor Filippo Colarossi, who purchased a school established in 1815 on the Île de la Cité and relocated the academy to 10 rue de la Grande Chaumière in the 1870s; the school closed in the 1930s, and Madame Colarossi destroyed the school's archives.

11 Cassirer had studied in Munich and was coeditor of *Simplicissimus* from 1896 to 1898. He opened a Berlin gallery in 1908 at 35 Viktoriastrasse to promote the work of Berlin Secession artists alongside that of their avant-garde colleagues in Paris.

12 Hunter later states that Hofmann knew of movements such as German Expressionism but "had no contact with them" (2002, 13).

13 The first exhibition of Synchromist painting was at the Neue Kunstsalon in Munich in June 1913; the second exhibition took place at the Bernheim-Jeune Gallery in Paris in October–November 1913.

14 Apollinaire first used the term *Orphism* in response to Kupka's work, at a lecture given at the 1912 Salon de la Section d'Or. Room 45 of the 1913 Salon des Indépendants featured the Orphist and Synchromist works of Delaunay, Terk, Kupka, Russell, and Macdonald-Wright.

15 Cheney's nephew Warren participated in the Munich school's summer sessions in Saint Tropez in the late 1920s. According to historian Dore Ashton, American painter Carl Holty "went all the way to Germany in 1925 to study with Hans Hofmann in order to satisfy his need for contact" with the modernist painting community (1972, 13).

16 Nevelson attended the school in 1931; Vytlacil, 1922–27; Jensen, 1926–27; and Paalen, 1927 (Hofmann student registries 1918–58).

17 A succession of artists taught classes at Hofmann's Munich school in his absence. The Munich school closed in 1936 (Dickey 2011, 111).

18 Former Hofmann students teaching in Minneapolis at the time

included Cameron Booth, Vaclav Vytlacil (who also taught at the Art Institute of Chicago and the Art Students League in New York City), John Haley (soon to join the faculty at Berkeley), and Edmund Kinzinger (Dickey 2011, 92).

19 First published by the university's press in 1943, Loran's book *Cézanne's Composition: Analysis of His Form, with Diagrams and Photographs of His Motifs* summarized the university's fine arts curricula. According to University of California, Berkeley history professor and Regional Oral History Office director Richard Cándida-Smith, the book was used for many years as a textbook for the department (2010).

20 Still taught at the California School of Fine Arts from 1946; Rothko, in the summers of 1947 and 1949; and Reinhardt, during the summer of 1950 (Sandler 1978, 35).

21 Still also studied with Hofmann student Vytlacil in the summer of 1946 before teaching in San Francisco (Dickey 2011, 241). Cheney's 1934 *Expressionism in Art* is listed in the bibliography of Still's master's thesis (Cheney 1948; Still 1935).

SIDEBAR 1

Modern Paint Technology in Germany

a A patent for aniline purple dye was obtained three years later by August Hofmann's student William Henry Perkin, who is credited improperly with the discovery (Ball 2001a, 210–13).

b Zinc (oxide) white was introduced in watercolor paints in 1834 and became commercially available as an artist's oil paint in the 1850s. Cadmium yellows and oranges were available shortly before Hofmann's birth, but cadmium red was not commercially available until 1910, and the now-familiar shades of cadmium red were probably not available until Bayer introduced them in 1919. Alizarin dyes were discovered in 1869; fluorescein dyes, in 1871; and rhodamine dyes, in 1887 (all manufactured by BASF). Diazo dyes entered commercial production in 1859; monoazo colors, in 1869 (a second group of monoazo colors was formulated in 1889). See Boxall 1978a, Ball 2001a, and Lomax and Learner 2006.

SIDEBAR 2

Modern Paint Technology in France

c "For the B quality this proportion may be up to 70%" (de Keijzer 2002, 42).

d Hansa yellows (also from the nitrogen-based "azo" or "arylide" color group) were introduced in Germany in 1911 and became popular throughout Europe by the 1920s. Diarylide yellows and oranges were introduced shortly after the arylide yellows (Lomax and Learner 2006, 109).

e Cadmium red was not available until 1910, although lighter shades of cadmium were being introduced in the years prior to the release of cadmium red. Scholar Philip Ball has since noted, "Since this was 1904, however, [Matisse] was presumably speaking of whatever pigment it was that George Field had investigated, whose orange tint would have resembled vermilion more closely than the deeper modern variety" (Ball 2001a, 303).

Chapter 2

EPIGRAPHS

Perls Gallery 1951; Rosenberg 1963a, 136.

22 The other two painters chosen by Seitz were Robert Motherwell and Mark Tobey. Seitz's dissertation was circulated in photocopied form among members of the arts community before its 1983 publication by Harvard University Press.

23 Hofmann taught at multiple schools, including his own; Gorky taught drawing at the Grand Central School for Art from 1926 to 1931; Rothko gave lessons at the Subjects of the Artist School; and de Kooning taught students privately.

24 Barr mentions Kandinsky's "abstract expressionism" in Museum of Modern Art catalogues in 1929, 1932, 1934, and 1936.

25 During this time the contemporary art division of Brandt's gallery was managed by Betty Parsons, who took over Brandt's lease and opened her own gallery at the same location when Brandt moved to England after World War II. Parsons represented Hofmann prior to his move to the gallery run by Samuel Kootz.

26 For Hofmann's return to painting, see Wolanin 2003.

27 Both oil paint and water-based materials appear in Hofmann's paintings from the 1930s and 1940s (Rockwell 1982). Works on canvas appear among the limited extant examples of Hofmann's painting from his years in Europe. Works on panel continue to appear into the early 1950s.

28 Seitz was curator of the Department of Painting and Sculpture at the Museum of Modern Art from 1960 to 1970.

29 Hofmann's painting *Spring*, for example, bears a date of 1940 but is dated in museum records to 1944–45, a period of similar extant works by other Abstract Expressionist artists. Hans Hofmann, *Spring*, 1944–45 (dated on reverse 1940), oil on wood, 28.5 × 35.7 cm (11¼ × 14⅛ in.). Gift of Mr. and Mrs. Peter A. Rübel, collection of the Museum of Modern Art.

30 Kline had a studio at 473 Sixth Avenue from 1944 to 1957; Reinhardt's studio was at 732 Broadway, between Waverly Place and Eighth Street; Ashton and Rosenberg lived in the Tenth Street neighborhood. For a complete list of Hofmann's addresses, see Dickey 2011, 163, 171, 184, 240.

31 Pollock and Krasner shared a studio at 46 East Eighth Street from 1935 to 1946.

32 The Waldorf Cafeteria was located at 390 West Eighth Street, at the corner of Sixth Avenue. In 1933 the Cedar Tavern was established in the neighborhood at 55 West Eighth Street, relocating to 24 University Place (at the corner of Eighth and University) in 1945. From 1937 onward, Hofmann's New York school was located at 52 West Eighth Street. The Subjects of the Artist School was founded in 1948 at 35 East Eighth Street. The Club was located at 39 East Eighth Street.

33 On May 20, 1950, twenty-eight abstract artists signed an open letter to the museum's president Roland L. Redmond, protesting the museum's jury selection for its acquisitions-based exhibition *American Art Today*. The letter was reprinted two days later on the front page of the *New York Times* under the heading "18 Painters Boycott Metropolitan; Charge 'Hostility to Advanced Art.'" The eighteen painters who signed the letter were Jimmy Ernst, Adolph Gottlieb, Robert Motherwell, William Baziotes, Hans Hofmann, Barnett Newman, Clyfford Still, Richard Pousette-Dart, Theodoros Stamos, Ad Reinhardt, Jackson Pollock, Mark Rothko, Bradley Walker Tomlin, Willem de Kooning, Hedda Sterne, James Brooks, Weldon Kees, and Fritz Bultman. The ten sculptors who signed the letter were Herbert Ferber, David Smith, Ibram Lassaw, Mary Callery, Day Schnabel, Seymour Lipton, Peter Grippe, Theodore Roszak, David Hare, and Louise Bourgeois. *Life* magazine sent photographer Nina Leen to capture the moment in a group portrait, which was published by the magazine in January 1951 with the caption "Irascible group of advanced artists led fight against show." The incomplete portrait depicts only fifteen of the painters and none of the sculptors, but it has become a de facto record of prominent Abstract Expressionist artists.

34 "I have had a very bad shock to find that I could no longer hold Hans Hofmann," Parsons wrote to collector Wright Ludington on March 10, 1947. "Sam Kootz, the crocodile, has gobbled him" (Parsons Gallery records, ca. 1920–91; Bystryn 1978).

35 Hofmann had previous solo exhibitions in San Francisco (California Palace of the Legion of Honor, August 1931) and New Orleans (Isaac Delgado Museum of Art, March 1941).

36 Hofmann's first show at Art of This Century ran from March 7 to 31, 1944. Greenberg cited the gallery's 1943–46 solo shows of works by Pollock, Hofmann, Baziotes, Motherwell, Rothko, and Still in "American-Type Painting" (1955). The opening exhibition of *Abstract and Surrealist Art in America* ran from November 29 to December 30, 1944, at the Mortimer Brandt Gallery. Hofmann was also included in Janis's earlier touring version of the exhibition, entitled *Abstract and Surrealist Art in the United States*. *Forty American Moderns* showed in December 1944; also exhibited: Milton Avery, Stuart Davis, Adolph Gottlieb, Morris Graves, Motherwell, Pollock, Richard Pousette-Dart, Rothko, and Mark Tobey. *A Problem for Critics* ran from May 14 to July 7, 1945; also exhibited: Hans Arp, Gorky, Gottlieb, Krasner, Masson, Roberto Matta, Miró, Picasso, Pollock, Pousette-Dart, Rothko, Charles Seliger, and Rufino Tamayo. *The Ideographic Picture* ran from January 20 to February 8, 1947; also exhibited: Newman, Reinhardt, Rothko, Theodoros Stamos, Hedda Sterne, and Still.

37 The March 19, 1951, symposium was entitled "How to Combine Architecture, Painting, and Sculpture."

38 This quotation refers to an oft-repeated exchange between Hofmann and Pollock in which Hofmann suggests that Pollock's painting would improve if he derived his subject matter from the natural world, and Pollock flippantly replied, "I am nature" (Krasner 1968).

39 A. E. Gallatin's Gallery of Living Art was originally located in a New York University study hall adjacent to Washington Square Park. The gallery was renamed the Museum of Living Art in 1936. The Whitney Studio was located at 8 West Eighth Street. It was renamed the Whitney Museum of American Art in 1931, moved to West Fifty-Fourth Street in 1954, to 945 Madison Avenue in 1966, and then to 99 Gansevoort Street in 2015. The Whitney's Eighth Street building became the home of the New York Studio School of Drawing, Painting, and Sculpture, founded by Hofmann student Mercedes Carles Matter.

40 Hofmann students who were founding members of the American Abstract Artists: Josef Albers, Rosalind Bengelsdorf, Ilya Bolotowsky, Harry Bowden, Byron Browne, George Cavallon, Burgoyne Diller, Ray Kaiser Eames, Carl Holty, Harry Holtzman, Leo Lances, Agnes Lyall, Mercedes Carles Matter, George McNeil, John Opper, Esphyr Slobodkina, Albert Swinden, Vaclav Vytlacil, and Wilfred Zogbaum (Hofmann student registries 1918–58). Although Hofmann was not a member of the American Abstract Artists, he was invited to address the organization, most notably on February 16, 1941, at a symposium on abstract art held during the *American Abstract Artists' Fifth Annual Exhibition*, presented at the Riverside Museum, New York (Lizza 1986, 111–12).

41 "I studied painting technique with Max Doerner [in the] Munich Academy.... And then I also went to the Hofmann School in the evenings—Hans Hofmann. He was there too" (Albers 1968).

42 Mercedes Carles (later Mercedes Carles Matter) began studying with Hofmann at the Art Students League. She was the daughter of American painter Arthur B. Carles, a friend from Hofmann's Paris years.

43 Hofmann students who were founding members of the Tenth Street cooperative galleries: Cecily Barth, Margaret Bartlett, Robert Beauchamp, Jacques Beckwith, James Billmyer, Nieves Billmyer, Barbara Forst, Miles Forst, William Freed, James Gahagan, Charles (Red) Grooms, Tom Hannan, Myrna Harrison, Robert Henry, Wolf Kahn, Allan Kaprow, John Kazann, John Krushenick, Nicholas Krushenick, Charles Littler, Marcia Marcus, Alvin Most, Haynes Ownby, Patricia Passlof, Don David, Peter Stander, Richard Stankiewicz (Hofmann student registries 1918–58).

44 Originally the WPA also employed foreign artists who had applied for permanent resident status, including Gorky and Rothko. De Kooning left WPA employment in 1937, when U.S. citizenship became a requirement. Artist employment programs available through the U.S. Department of the Treasury: Public Works of Art Project (December 1933–June 1934, funded by the Civil Works Administration), Section of Painting and Sculpture (October 1934–October 1938, then renamed the Section of Fine Arts; oversight trans-ferred to the new Federal Works Agency in 1939, which administered the program until June 1943). Artist employment available through the WPA: Federal Art Project (August 1935–September 1939, when oversight was transferred to the Federal Works Agency, which administered the program until March 1942), Treasury Relief Art Project (July 1935–June 1939, funded by the WPA and administered by the Treasury Department).

45 At the time Hunter was chief curator and acting director of the Minneapolis Institute of Arts.

46 The series—sometimes referred to as "X Paints a Picture"—ran from 1949 through 1966, with the majority of articles appearing in the 1950s. Abstract Expressionists featured in the series included Elaine de Kooning (December 1960), Willem de Kooning (March 1953), Gottlieb (March 1955), Hofmann (February 1950), Kline (December 1952), Pollock (May 1951), and Reinhardt (March 1965).

47 Still and Rothko also controlled the conditions under which their works could be viewed, with Rothko often choosing the wall color and lighting design for the exhibition space (Fer 2005).

48 Ad Reinhardt, in a letter to Sam Hunter in the summer of 1966 (Rowell 1980, 21).

49 "This full-bodied performance made literal the implied kinesthetic response to a Pollock pouring.... It also prefigures the 'living brush' technique that Yves Klein directed around 1960" (Kachur 2007, 155).

50 The supply list can be tentatively dated between the original 1934 publication of Cheney's *Expressionism in Art* (included on the list; Cheney 1948) and Rose Kuper's studies with Hofmann in the late 1930s (Kuper collection, 1932–64).

51 According to Rosenberg, "The American Action Painters" (Rosenberg 1952) was initially intended to list several artists by name, but the names were removed when the article was published for a foreign audience. "I was going to put some names in... like Pollock, Hofmann, and de Kooning, but then I didn't want to put in too many names because it would confuse the issue.... So I took all names out" (Rosenberg 1972).

SIDEBAR 3

Modern Paint Technology in the United States

f According to conservator Ralph Mayer, Keim's technique was also attempted for decorations in the New York City subway system ("Subway murals, 1937–38," Mayer papers, [ca. 1920]–64).

g DuPont's pyroxylin brand name was thereafter associated with Siqueiros through his nickname "Il Duco."

h Porter's reference to "Glyptal thinner" likely refers to a proprietary solvent mixture appropriate for use with paints containing alkyds.

i Duco is often associated with Pollock; according to Helen Harrison, Eugene V. and Clare E. Thaw Director of the Pollock-Krasner House and Study Center in East Hampton, New York, recent analysis of paints in the artist's studio shows more prevalent use of house paint, floor paint, and radiator paint by F.W. Devoe and C.T. Raynolds Company and by Pittsburgh Paints (Pittsburgh Plate Glass Co.) (Harrison 2014).

j The origin of that sample has been attributed to several different artists working on murals for the Federal Art Project (Bocour 1978; Lodge 1988, 125).

Chapter 3

EPIGRAPHS

Hofmann 1931, 5; 1948, 46.

52 Hofmann previously worked with architects José Luis Sert and Paul Lester Weiner on mural designs for a church in Chimbote, Peru, an unrealized project in conjunction with the October 1950 Kootz exhibition *The Muralist and the Modern Architect.* Kootz paired artists with architects to work on site-specific designs. Other pairings included William Baziotes and Philip Johnson, Adolph Gottlieb and Marcel Breuer, David Hare and Frederick Kiesler, and Robert Motherwell and the Architect's Collaborative (a firm led by Walter Gropius).

53 Other students working on the mural project included William Freed and Robert Fisher. Similar colored papers were used for Hofmann's 1958 design for a mural at the New York School of Printing (now the High School of Graphic Communication Arts) at 439 West Forty-Ninth Street (Dickey 2002, 274).

54 Accounts differ of the exact number of works stored by Hofmann and Wolfegg. At least two dozen works are listed in a letter to Gabriele Münter dated June 3, 1927 (Dickey 2011, 73). The nature-based origins of Kandinsky's nonobjective paintings are discussed in Rose-Carol Washton Long's *Kandinsky: The Development of an Abstract Style* (1980).

55 The tilted color plane in Hofmann's *Imperium in Imperio* (1964, collection of the University of California, Berkeley Art Museum and Pacific Film Archive) is a rare example of an overtly figured plane that is not positioned parallel to a canvas border.

56 Hofmann received honorary doctor of fine arts degrees from three American institutions: Dartmouth College in New Hampshire (1962), the University of California, Berkeley (1964), and Pratt Institute in New York (1965). He was awarded honorary membership in the Akademie der Bildenden Künst in 1962 and became a member of the National Institute of Art and Letters (now the American Academy of Arts and Letters) in 1964.

57 Einstein was awarded a 1921 Nobel Prize for his discovery of the law of the photoelectric effect, central to the study of quantum physics (he received the award in 1922). Hofmann discussed the fourth dimension in his 1930 *Art Digest* essay "Art in America" (1930a).

58 Portions of the unedited interview included in this book were provided by Feinstein's son Sascha.

59 "Notes on Technique," Goodman 1986, 113–16. The information is drawn from Lillian Olinsey Kiesler's January 9, 1983, interview with Goodman (Kiesler 1983), a section on Hofmann in Katherine Kuh's 1962 book *The Artist's Voice: Talks with Seventeen Modern Artists*, and a 1982 unpublished conservation survey by Tony Rockwell on the Hofmann Collection at the University of California, Berkeley Art Museum and Pacific Film Archive.

60 De Kooning's article mentions six versions of the painting, while Hofmann's painting ledger for the period lists versions two through five and a "final" (likely version six, now in the collection of the Neuberger Museum of Art, Purchase College, State University of New York). Hofmann's ledger of works consigned to Kootz likewise mentions only versions two through five. The whereabouts of version one are unknown.

61 Feinstein notes that his filming of Hofmann was prompted by de Kooning's article (Feinstein 2008, 21). Hans Hofmann, *The Window*, 1950, oil on canvas, 121.9 × 91.8 cm (48 × 36⅛ in.). Gift of Mr. and Mrs. Roy R. Neuberger, 1951, collection of the Metropolitan Museum of Art, New York.

62 Permalba is a white pigment oil color manufactured by the Philadelphia art materials manufacturer F. Weber and Company. The name Permalba is commonly associated with a blend of titanium and zinc white pigments.

63 Filmed clips of Hofmann at work also appear in Madeline Amgott's documentary *Hans Hofmann: Artist/Teacher, Teacher/Artist* (2003) and Warren Forma's film *The Americans: Three East Coast Artists at Work* (1963), although no substantive information regarding Hofmann's methods or materials is provided. Forma's film also features painters Milton Avery and Jack Tworkov.

64 Interview with Sam Feinstein conducted in December 1989 by Sascha Feinstein (Feinstein 1989); courtesy of Sascha Feinstein.

65 Both notations include measurements for stretcher bars. Another note reads, "For strecher [*sic*]: Clinton Schwab & Co., Manf. Of Moldings, 29–33 Lowell Str. Arlington [illeg.], Mass."

66 Rosenthal as a G.I. Bill supplier is mentioned.

67 Some clarification in terminology is necessary here. Once Hofmann switched from wood panels to paintings on canvas, the term *gesso* no longer applied to his ground layer materials. Gesso more exactly refers to the calcium sulfate ground layer material applied to wood panels (*gesso* is the Italian word for gypsum), but the term was commonly misapplied by twentieth-century writers and artists to the oil-based (and later, after the 1940s, acrylic) ground layer materials used to prepare paintings on canvas (*canvas* is another loose term meant to apply to linen fabrics but commonly used in reference to both cotton and linen materials).

68 Artists' priming layers should not be confused with the commercial primer paints used to seal interior and exterior walls before painting (thereby functioning as ground layers).

69 Hans Hofmann, *Magenta and Blue*, 1950, oil on canvas, 122.1 × 147.5 cm (48 1/16 × 58 1/16 in.). Purchase, collection of the Whitney Museum of American Art.

70 Tempera paints are mentioned in two photo captions in de Kooning's 1950 article: "his materials for tempera painting" and "working on tempera sketches as a 'warm-up' prior to painting in oils" (39). There is no other mention of traditional egg tempera paints in Hofmann's studio practice; this is another loose application of terminology that likely refers to the watercolors and water-based gouache or casein paints Hofmann used for his works on paper. Tempera painting materials are not mentioned in the main article text.

71 A combination of German, French, and English is frequently seen in documents from Hofmann's early years in the United States (Hofmann papers, [ca. 1904]–78).

72 Permalba appeared in artists' paints in 1920; the range of cadmium colors, from 1880 to 1909; cerulean blue, 1862; cobalt violet, 1859; alizarin crimson, 1869 (Boxall 1978a; Ball 2001a).

73 Duplicate copies of the supply list were found in the Kiesler papers ([ca. 1910]–2003) and in the Rose Kuper collection of Hans Hofmann miscellany (1932–64). The reverse of the supply list in the Hofmann papers contains a partial draft of an October 31, 1944, letter from Hofmann to Federal Security Agency attorney Harry N. Rosenfield inquiring about application of the G.I. Bill of Rights to registrants of his school, which suggests that the supply list was still in use in the mid-1940s.

74 Hofmann's use of the term *medium* is again an example of loose terminology on the part of the artist, as oil and beeswax are the only true mediums in that list (the rest are diluents).

75 When Koch studied with Hofmann in Provincetown, he also worked

at an art store owned by painter Jim Forsberg called "the studio shop," where he frequently sold the components for this medium (Koch 2012). See also Porter 1953, 73; Campbell 1954, 35.

76 Dickey was involved in the gathering and review of the artist's papers and confirmed that the handwriting on the methacrylate varnish note is not in keeping with the recognized writing of Hofmann, any member of his family, or his students.

77 It was common practice in that period to varnish works before they were shipped. Some of Hofmann's works may have been varnished in that manner, although no records exist regarding this practice in relation to the works held by the Hofmann Trust (Yohe 2013).

78 Hofmann's personal papers were donated to the Smithsonian Institution's Archives of American Art in 1997.

Chapter 4

EPIGRAPHS

Hofmann 1959, 10; Hofmann papers, [ca. 1904]–78.

79 Hofmann's political views are not discussed in interviews with the artist, although his affiliation with modern art and liberal activists clearly was not in line with Nazi politics. Hofmann became an American citizen in 1941 (Friedel 1997, 9).

80 Hofmann's 1963 retrospective exhibition ran at the Museum of Modern Art from September 11 to November 28 and toured to five national and seven international venues.

81 Selz was curator of the Department of Painting and Sculpture at the Museum of Modern Art from 1958 to 1965.

82 The period inclusion of oil-soluble additives and acrylic plasticizing agents including alkyd resins, nitrocellulose, and vinyl polymers must also be considered (Weldon 2001; Learner 2004).

83 Duco switched from a cellulose nitrate base to an alkyd base in the mid-1930s (Standeven 2003, 9; Lake, Ordonez, and Schilling 2004, 140). It is unclear whether artists knew of such shifts in formulation.

84 Lake noted that "reports that [Willem] de Kooning's unusual medium included mayonnaise may be an extrapolation of the vivid descriptions of the artist's temporary emulsion as having the 'consistency of heavy sauces' and 'a good mayonnaise'" (Lake, Lomax, and Schilling 1999, 383). No evidence of egg protein was found during Lake's analysis of the paint.

85 According to Helen Harrison, the reference in Barbara Rose's published interview with Lee Krasner (Rose 1980) to special paints from DuPont is incorrect (Harrison 2014). In the corrected interview transcript (p. 5), DuPont is crossed out and Devoe is added (http://www.aaa.si.edu/collections/container/viewer/Barbara-Ros-Interview-286262).

86 Tallent's survey included two Hofmann works from the collection of the Albright-Knox Art Gallery. The two works, *Exuberance* (1955) and *Sommernachtstraum* (1957), are both included in the group of paintings selected for this study.

87 Tallent's identification of Hofmann's materials was based on visual examination of the paintings and optical microscopy of selected paint samples. No further analysis was performed on the samples taken for analysis, and the samples were subsequently discarded (Tallent 2005).

88 The ground layers and compositional paints on those palettes have been reexamined in this research.

89 None of the Smithsonian Institution paintings are included in the current study.

90 Here Rockwell uses the term *gesso* in reference to the ground layers on Hofmann's wood panels. When discussing the preparatory materials Hofmann used for works on canvas, Rockwell switches terminology and uses the term *ground*.

91 Noted by Tallent (1988) during examinations in collaboration with Robert Lodge, then conservator of modern paintings at the Intermuseum Conservation Association, Oberlin, Ohio (now in Cleveland).

92 M536-45 and M536-03, collection of the Renate, Hans and Maria Hofmann Trust. Both palettes are included in the group selected for this study.

93 Paint can lid, diameter 15.2 cm (6 in.), M536-49, collection of the Renate, Hans and Maria Hofmann Trust.

94 Tallent (1988) noted a fluorescence consistent with other alkyd paints within a rectangle of white paint in the Hofmann Collection painting *Lucidus Ordo*, 1962, oil on canvas, 213.7 × 198.1 cm (84⅛ × 78 in.). Bequest of the artist, collection of the University of California, Berkeley Art Museum and Pacific Film Archive.

95 *Radiance*, 1956, oil on canvas, 64.1 × 76.8 cm (25¼ × 30¼ in.); gift of Joseph H. Hirshhorn. *Flowering Swamp*, 1957, oil on canvas, 122 × 91.5 cm (48⅛ × 36⅛ in.); gift of the Joseph H. Hirshhorn Foundation. *Prelude of Spring*, 1958, oil on canvas, 127.6 × 214 cm (50¼ × 84¼ in.); gift of the Joseph H. Hirshhorn Foundation. *Oceanic*, 1958, oil on canvas, 152.7 × 121.9 cm (60⅛ × 48 in.); gift of the Joseph H. Hirshhorn Foundation. *To JFK: A Thousand Roots Did Die with Thee*, 1963, oil and enamel on canvas, 152.4 × 182.8 cm (60 × 72 in.); gift of Mr. and Mrs. Leigh B. Block. All paintings, collection of the Hirshhorn Museum and Sculpture Garden, Smithsonian Institution, Washington, DC.

96 Surveys of the Hofmann Collection were performed in 1974, 1979, 1982, 1986, 1988, 1989, 1991, 1999, 2003, and 2009 (BAM/PFA records 1966–present; SFMOMA 1966–present).

97 As visual evidence of Hofmann reworking a composition, Arthur Coleman Danto (1990) juxtaposes a work signed by Hofmann in 1959 with a 1960 photograph by Arnold Newman of Hofmann revisiting the work; several colored papers are pinned to its surface.

98 As of my examination of the paintings in December 2011. Sticky paint was also observed on one Trust palette during my visit that month.

99 Both dry and liquid exudates are sometimes loosely referred to by the crystalline exudation descriptive *efflorescence*; this should not be confused with the glowing appearance of some exudates under ultraviolet illumination—not all efflorescence fluoresces.

100 Records alternately refer to Dinnean as curator of conservation and as curator of the Hofmann Collection.

101 Dinnean's note appears next to the entry for a painting designated for treatment on November 10, 1972.

102 Information compiled from the University of California, Berkeley Art Museum and Pacific Film Archive curatorial files (Hofmann Collection treatment and administration records); San Francisco Museum of Modern Art conservation files (Elise S. Haas Conservation Department records, including Rockwell's 1982 survey report); Intermuseum Conservation Association records (treatment files for study group paintings from the Albright-Knox Art Gallery and Memorial Art Gallery and Tallent's survey records of the Berkeley collection, plus her personal notes related to the survey and summary information communicated to the author [Tallent 1988, 2005]); and the conservation records of the Albright-Knox Art Gallery, the Memorial Art Gallery, and the Museum of Modern Art.

103 A publication detailing the treatment was preceded by a December 8, 2001, symposium entitled "Restoration of *Cathedra*—Considering the Restoration of a Monochrome Painting" (Stedelijk Museum Amsterdam 2002).

104 Photographs of the canvas versos taken prior to lining were not accessible for this study.

105 Poly(vinyl acetate) resins with varying glass transition temperatures are usually mixed to customize handling and structural characteristics. Poly(vinyl acetate) formulations are designated by the coding AYAA, AYAB, AYAC, and so on. AYAA has the lowest molecular

weight; AYAF is the highest molecular weight formulation commonly used for conservation purposes.

106 I have substituted *modern* for *Cubist* in the quote, as a footnote to Richardson's comment reads, "For the purposes of this article I have restricted myself to the Cubist paintings of Braque and Picasso. However, my strictures equally apply to the insensitive way many other twentieth-century artists' work has been treated" (1983, 32). Selected responses from conservators to Richardson's comments about both varnishing and lining were published in "Letters," *New York Review of Books* 30 (17) (November 10, 1983): 60–64. Gustav Berger's response was entitled "Saving the Cubists."

107 Analysis by conservators at the Solomon R. Guggenheim Museum revealed that early application of a synthetic varnish to one of Ad Reinhardt's *Black Series* paintings may have protected the original paint—although not the original paint surface—from subsequent layers of extensive overpaint (Stringari, Pratt, and McGlinchey 2004).

108 "I never glaze or varnish a painting" (*Magenta and Blue* typescript, Hofmann papers, [ca. 1904]–78).

109 The memo is copied to "Peter, David, and Joan." The identity of David could not be definitively ascertained.

110 Dinnean's penciled note reads, "Not controversial among informed persons" ("To: Larry / From: Brenda," May 25, 1972, BAM/PFA records 1966–present).

111 Dinnean's penciled side note reads, "Erle states unequivocally that they should be varnished" ("To: Larry / From: Brenda," May 25, 1972, BAM/PFA records 1966–present).

112 Goldreyer initially states that all eight paintings were "sealed with B.M.P." (BAM/PFA records 1966–present, "Nov. 72"); a request for more information may have prompted the undated note, although no such request appears in the museum records.

Chapter 5

113 GC-MS analysis was performed at the National Gallery of Art in January and March 2012; SEM-EDS analysis was performed at the Museum Conservation Institute, Smithsonian Institution (MCI), in March and April 2012; FTIR analysis was performed at MCI in December 2012 and in January and March 2013; XRD analysis was performed at MCI in September 2013.

114 Plasticizer transfer from some storage containers can give false positive results for alkyd paint in GC-MS and FTIR analyses.

115 Alkyd resins, nitrocellulose, and vinyl polymers were used as modifying agents in some 1950s-era acrylic paint formulations (Learner 2000).

116 Zinc white alkyd formulations were unstable and not commercially viable during Hofmann's career, and given the artist's preference for zinc oxide pigment, a zinc-lacking paint in Hofmann's work is a good indicator of an alkyd paint formulation.

117 EDS analysis of visible inclusions—such as mordant materials for dye-based pigments—was also performed but not included in the analysis tally provided in table 5 (see appendix 1).

118 A label from a paint can provided by the Benjamin Moore archives and date-stamped June 9, 1959, lists the paint's inorganic components as 49.9% titanium calcium, 31.7% calcium carbonate, and 18.4% silicates (Rogala 2005).

119 No compositional white paints were present in *Heraldic Call* (1962) or on Trust palette M536-03.

120 Sampling from palettes was limited to priming layers and components of mixed colors. Other colors in consistent late-career use may also appear on the palettes.

121 The high levels of barium found through elemental analysis suggest barium sulfate over lithopone as an extender, although the primary point is that an extender other than calcium sulfate is present in this sample.

122 The blue in *Heraldic Call* (1962) was not sampled but is a visual match to other paints in Hofmann's late-career oeuvre identified as ultramarine blue.

123 Some paints that appear black were later identified as phthalocyanine green.

124 As noted in table 3 (see appendix 1), conservation records indicate that AYAA appears in conjunction with dry pigments used for inpainting, while AYAC is a component of the inpainting material Mowilith 20.

125 A match to the International Center for Diffraction Data (ICDD) libraries reference pattern for $CdS_{0.54}Se_{0.46}$.

126 A match to the ICDD libraries reference pattern for $CdS_{0.42}Se_{0.58}$.

127 Contemporaneous industrial pigment catalogues courtesy of Georg Kremer.

128 A match to the ICDD libraries reference pattern for $Na_{6.36}Ca_{1.52}(AlSiO_4)_6(SO_4)_{0.84}S_{1.544}$.

129 A match to the ICDD libraries reference pattern for $(Cd_{0.65}Mn_{0.35})S$.

130 This unique resonance has since been exploited for use in nanotechnology. See M. Ragam et al., "Localized Vibrational Mode in Manganese-Doped Zinc Sulphide and Cadmium Sulphide Nanoparticles," *Defect and Diffusion Forum* 318 (July 2011): 11–21.

Chapter 6

EPIGRAPH
Clark 1994, 26.

131 Alkyds are also oil based but distinctly different in appearance and handling from artist's oil paints.

132 Yellow paint that is a visual match to other cadmium yellows in Hofmann's work is visible on *Heraldic Call* (1962), but it was not sampled.

133 Ball (2001a, 168–96) identifies the following colors as materials typically used in Impressionist paintings: zinc white, lead white, lemon yellow (barium chromate), chrome yellow (lead chromate), cadmium yellow, Naples yellow (lead antimonite), yellow ocher, chrome orange (basic lead chromate), vermilion, red ocher, natural madder lake, crimson (cochineal) lake, Scheele's green (copper arsenite), emerald green (copper acetoarsenite), viridian (hydrated chromic oxide), chrome green (a mixture of Prussian blue and chrome yellow), cerulean blue (cobalt stannate), cobalt blue (cobalt aluminate), artificial ultramarine, and ivory black (bone black).

134 Chevreul's *De la Loi du Contraste Simultané des couleurs* was first published in 1839 (Paris: Pitois-Levrault); the first English translation—entitled *The Principles of Harmony and Contrast of Colors and Their Applications to the Arts of Painting, Decoration of Buildings, Mosaic Work, Tapestry and Carpet Weaving, Calico Printing, Dress, Paper Staining, Printing, Illumination, Landscape and Flower Gardening etc.*—appeared in 1854 (London: Longman and Co.). Rood's *Modern Chromatics, with Applications to Art and Industry* was first published in 1879 (New York: D. Appleton and Co.), with German and French translations appearing in 1880 (Leipzig: Brockhaus) and 1881 (Paris: G. Baillière).

135 Industrial enamel paints have been found in works produced by Picasso and Francis Picabia in the early 1920s. See *Journal of the American Institute for Conservation* 52 (3–4) (August and November 2013), postprints from the May 2011 symposium "From Can to Canvas: Early Uses of House Paints by Picasso and His Contemporaries in the First Half of the 20th Century."

136 Ellsworth Kelly lived in Paris for six years; he had moved back to New York by 1954.

137 Phthalocyanine-based pigments entered industrial production by E.I. du Pont de Nemours and Company in the United States in 1936; phthalocyanine blue was introduced in Winsor and Newton's artists' paints in 1937, and phthalocyanine-based green appeared in Winsor and Newton's artists' paints in 1939; phthalocyanine-based colors appeared in Talens's Rembrandt paint series in 1940 (de Keijzer 2002, 51–52).

138 The quinacridone-based rose-colored paints mentioned by Seckler 1953 (Rembrandt Rose [Davis]) and Campbell 1954 (Shiva Rose Red [Ferren]) are among several alizarin alternatives that were introduced in the late 1950s and early 1960s (de Keijzer 2002, 52; Lomax 2005).

139 A companion set of demonstration materials related to Albers's theories was also produced.

140 Muddied tones were purposeful, as Lillian Kiesler and others recount Hofmann's fastidious cleaning practices (Kiesler 1983, 4–5; Dickey 2011, 310).

141 Localized voids were not visible in *Combinable Wall I and II* and *Silent Night*, but they may exist.

142 This lecture is the starting point for Clark's essay "In Defense of Abstract Expressionism," published in *October* 69 (Summer 1994) and in a revised format in *Farewell to an Idea: Episodes from a History of Modernism* (1999).

143 According to Erle Loran, Hofmann always sized his canvases. "Notes from one hour interview with Erle Loran, June 1987," courtesy of Carolyn Tallent (Tallent 1987).

144 On the following page, Toch notes that "the low price of benzine in America offers a great temptation for its unlimited use" (1925, 270).

145 Tallent (1988) also notes that benzine was used to test the solubility of exudates.

Conclusion

EPIGRAPHS

Shiff 1987, 95; Seitz 1955, 29.

146 According to Lake (1999, 395–96), de Kooning was a close friend of sculptor David Smith, who comanaged an experimental materials workshop through which his Abstract Expressionist colleagues had access to new art-making materials.

147 As of 2015, no paintings from the Hofmann Collection have been sold by the University of California.

148 Some short-lived mid-twentieth-century modifiers may be used to determine the date of manufacture—tin and formaldehyde biocides, for example; alkyd plasticizers in poly(vinyl acetate) formulations; and the post–World War II use of surplus styrene-butadiene rubber as a plasticizing agent. Period additives can also sometimes function as oil-length markers (Learner 2004, 18–19).

149 Greenberg (1939) and Wolfe (*The Painted Word* [New York: Farrar, Straus and Giroux, 1975]) primarily associate Abstract Expressionism's colors with the influence of the bourgeoisie, while Clark (1994) contends that the "vulgarity" of the bourgeois palette is in itself a modern concept.

References

ABBREVIATIONS

AAA
Archives of American Art, Smithsonian Institution
BAM/PFA
University of California, Berkeley Art Museum and Pacific Film Archive
SFMOMA
San Francisco Museum of Modern Art

Agee, William C. 2012. "Hans Hofmann: Art Like Life Is Real." In *Hans Hofmann: Art Like Life Is Real,* 5–13. New York: Ameringer McEnery Yohe.

Albers, Josef. 1968. Oral history interview by Sevim Fesci, June 22–July 5. AAA.

———. 1975. *Interaction of Color.* New Haven and London: Yale University Press. Originally published in 1963.

Amgott, Madeline. 2003. *Hans Hofmann: Artist/Teacher, Teacher/Artist.* Film (U.S. premiere, June 2003 [PBS]). New York: Amgott Productions.

Anfam, David. 1990. *Abstract Expressionism.* London: Thames and Hudson.

Art Digest. 1931. "Hofmann at Chouinard." *Art Digest* 5 (14): 29.

Ashton, Dore. 1972. *The New York School: A Cultural Reckoning.* New York: Viking Press.

———. 1995. "A Straggler's View of Gorky." In *Arshile Gorky: The Breakthrough Years,* edited by Michael Auping, 39–61. Fort Worth, TX: Modern Art Museum of Fort Worth in association with Rizzoli.

Auping, Michael. 1987. "Beyond the Sublime." In *Abstract Expressionism: The Critical Developments,* edited by Michael Auping, 146–66. New York: Harry N. Abrams in association with Albright-Knox Art Gallery.

Ball, Philip. 2001a. *Bright Earth: Art and the Invention of Color.* New York: Farrar, Straus and Giroux.

———. 2001b. "The Making of Cézanne's Palette." *Helix* 10 (2): 34–41.

BAM/PFA records [University of California, Berkeley Art Museum and Pacific Film Archive records]. 1966–present. Unpublished records related to the conservation of Hofmann works in the collection, including information related to conservation of works prior to museum acquisition.

Bannard, Walter D. 1969. "Hofmann's Rectangles." *Artforum* 7 (10): 38–41.

Bentley, J., and Gerald P. A. Turner. 1998. *Introduction to Paint Chemistry and Principles of Paint Technology.* London and New York: Chapman and Hall.

Berger, Gustav A., and Orrin H. Riley. 1971. "New Developments in the Conservation of Works of Art." *Art Journal* 31 (1): 77–86.

Blumenroth, Diana, Stefan Zumbühl, Nadim C. Sherrer, and Wolfgang Müller. 2014. "Sensitivity of Modern Oil Paints to Solvents: Effects on Synthetic Organic Pigments." In *Issues in Contemporary Oil Paint,* edited by Klaas Jan van den Berg, Aviva Burnstock, Matthijs de Keijzer, Jay Krueger, Thomas J. S. Learner, Alberto de Tagle, and Gunnar Heydenreich, 351–62. Amsterdam and Amersfoort: Netherlands Cultural Heritage Agency.

Bocour, Leonard. 1978. Oral history interview by Paul Cummings, June 8. AAA.

———. 1985. "Interview with Leonard Bocour." Morris Louis and Morris Louis Estate papers, 1910s–2007, bulk 1965–2000, AAA.

Bois, Yve-Alain. 2012. "Pablo Picasso: The Cadaqués Experiment." In *Inventing Abstraction, 1910–1925: How a Radical Idea Changed Modern Art,* edited by Leah Dickerman, 40–43. New York: Museum of Modern Art.

Bois, Yve-Alain, and Rosalind Krauss. 2000. *Formless: A User's Guide.* Cambridge, MA: MIT Press.

Boon, Jaap J., and Frank G. Hoogland. 2014. "Investigating Fluidizing Dripping Pink Commercial Paint on Van Hemert's *Seven*-Series Works from 1990–1995." In *Issues in Contemporary Oil Paint,* edited by Klaas Jan van den Berg, Aviva Burnstock, Matthijs de Keijzer, Jay Krueger, Thomas J. S. Learner, Alberto de Tagle, and Gunnar Heydenreich, 227–46. Amsterdam and Amersfoort: Netherlands Cultural Heritage Agency.

Boxall, J. 1978a. "A History of Paint Technology, Part Three: Mid-19th Century to 20th Century." *Paint Manufacture* 48 (6): 25–30.

———. 1978b. "A History of Paint Technology, Part Four: The 20th Century." *Paint Manufacture* 48 (7): 18–23.

Bradley, Morton C. 1950. *The Treatment of Pictures.* Cambridge, MA: Art Technology.

Brown, Julia. 1998. "A Conversation: Helen Frankenthaler with Julia Brown, Spring–Fall 1997, Connecticut and New York City." In *After "Mountains and Sea": Frankenthaler, 1956–1959,* 29–49. New York: Solomon R. Guggenheim Foundation.

Browne, Rosalind B. 1968. Oral history interview by Irving Sandler, January 29. AAA.

Buettner, Stewart. 1981. *American Art Theory, 1945–1970.* Ann Arbor, MI: UMI Research Press.

Bultman, Fritz (moderator). 1978. "Students Talk about Hofmann as Teacher." "Artists Talk on Art" panel discussion, March 17. Transcript, http://www.atoa.org/3-17-78.htm. Panelists included Lillian Olinsey Kiesler, Bultman, Nell Blaine, Jim Gahagan, Cynthia Goodman, George McNeill, and Selina Trieff.

Burnstock, Aviva, Klaas Jan van den Berg, Suzan de Groot, and Louise Wijnberg. 2007. "An Investigation of Water-Sensitive Oil Paints in Twentieth-Century Paintings." In *Modern Paints Uncovered: Proceedings from the Modern Paints Uncovered Symposium at the Tate Modern, London, May 16–19, 2006,* edited by Thomas J. S. Learner, Patricia Smithen, Jay W. Krueger, and Michael R. Schilling, 177–88. Los Angeles: Getty Conservation Institute.

Bustin, Mary, and Tom Caley, eds. 2003. *Alternatives to Lining: Structural Treatment of Paintings on Canvas without Lining. Preprints of a Conference Held Jointly by the British Association of Paintings Conservator-Restorers and the United Kingdom Institute for Conservation Paintings Section on 19 September 2003.* London: United Kingdom Institute for Conservation of Historic and Artistic Works.

Bystryn, Marcia. 1978. "Art Galleries as Gatekeepers: The Case of the Abstract Expressionists." *Social Research* 45 (2): 390–408.

Campbell, Lawrence. 1954. "Ferren Paints a Picture." *ARTnews* 52 (10): 34–37, 52–53.

Cándida-Smith, Richard. 2010. Personal communication with the author. December 1.

Carmean, E. A., Jr. 1978. *American Art at Mid-Century: The Subjects of the Artist.* Washington, DC: National Gallery of Art.

Caws, Mary Anne. 2003. *Robert Motherwell: With Pen and Brush.* London: Reaktion Books.

Cheney, Sheldon. 1924. *A Primer of Modern Art.* New York: Boni and Liveright.

———. 1948. *Expressionism in Art.* New York: Liveright Publishing. Originally published in 1934.

Chipp, Herschel B. 1957. "Diebenkorn Paints a Picture." *ARTnews* 56 (3): 44–47, 54–55.

Cho, Jennifer M. 1993. "Hans Hofmann and Joseph Albers: The Significance of Their Examples as Artist-Teachers." EdD diss., Columbia University Teachers College.

Claretie, Jules. 1881. *La Vie à Paris, 1881.* Paris: Victor-Havard.

Clark, Timothy J. 1992. Transcript of a lecture delivered in Hofmann Collection exhibition. Museum Education Department archives, University of California, BAM/PFA.

———. 1994. "In Defense of Abstract Expressionism." *October* 69 (Summer): 22–48.

———. 1999. *Farewell to an Idea: Episodes from a History of Modernism.* New Haven and London: Yale University Press.

Coates, Robert. 1946. "The Art Galleries: At Home and Abroad." *New Yorker* 22 (7): 83.

Coddington, James. 1999. "No Chaos Damn It." In *Jackson Pollock: New Approaches,* edited by Kirk Varnedoe and Pepe Karmel, 101–16. New York: Museum of Modern Art.

Colour Index. 1957. Yorkshire, England, and Lowell, MA: Society of Dyers and Colourists in association with American Association of Textile Chemists and Colorists.

Cooper, Anna, Aviva Burnstock, Klaas Jan van den Berg, and Bronwyn Ormsby. 2014. "Water-Sensitive Oil Paints in the 20th Century: A Study of the Distribution of Water-Soluble Degradation Products in Modern Oil Paint Films." In *Issues in Contemporary Oil Paint,* edited by Klaas Jan van den Berg, Aviva Burnstock, Matthijs de Keijzer, Jay Krueger, Thomas J. S. Learner, Alberto de Tagle, and Gunnar Heydenreich, 295–310. Amsterdam and Amersfoort: Netherlands Cultural Heritage Agency.

Corbeil, Marie-Claude, Jean-Pierre Charland, and Elizabeth A. Moffatt. 2002. "The Characterization of Cobalt Violet Pigments." *Studies in Conservation* 47 (4): 237–49.

Cranmer, Dana. 1987a. "Ephemeral Paintings on 'Permanent View': The Accelerated Ageing of Mark Rothko's Paintings." In *Preprints from the 8th ICOM Committee on Conservation Triennial Meeting in Sydney, Australia, 6–11 September 1987,* edited by Kirsten Grimstad, 283–85. London: James and James.

———. 1987b. "Painting Materials and Techniques of Mark Rothko: Consequences of an Unorthodox Approach." In *Mark Rothko: 1903–1970,* 189–97. London: Tate Gallery.

Crook, Jo, and Thomas J. S. Learner. 2000. *The Impact of Modern Paints.* London: Tate Gallery.

Crow, Thomas. 2005. "The Marginal Difference in Rothko's Abstraction." In *Seeing Rothko,* edited by Glenn Phillips and Thomas Crow, 25–39. Los Angeles: Getty Research Institute.

Danto, Arthur C. 1990. "Hans Hofmann." *Nation* 251 (7): 248–51.

de Antonio, Emile, and Mitch Tuchman. 1984. *Painters Painting: A History of American Modernism in the Words of Those Who Created It.* New York: Abbeville Press. Book based on transcript of de Antonio's 1972 film *Painters Painting: A Candid History of the Modern Art Scene.*

de Duve, Thierry. 1996. "The Monochrome and the Blank Canvas." In *Kant after Duchamp,* 199–280. Cambridge, MA: MIT Press. Adapted from materials originally written in 1986 and published in *Reconstructing Modernism,* edited by Serge Guilbaut, 244–310. Cambridge, MA: MIT Press, 1990.

de Keijzer, Matthijs. 2002. "The History of Modern Synthetic Inorganic and Organic Artists' Pigments." In *Contributions to Conservation: Research in Conservation at the Netherlands Institute for Cultural Heritage,* edited by Jaap A. Mosk and Norman H. Tennant, 42–54. Amsterdam: ICN.

de Kooning, Elaine. 1950. "Hans Hofmann Paints a Picture." *ARTnews* 48 (10): 38–41, 58–59.

De Maria, Walter. 1972. Oral history interview by Paul Cummings, October 4. AAA.

Diamonstein, Barbaralee. 1979. Interviews conducted at the New School of Social Research, New York. In *Inside New York's Art World,* 199–210. New York: Rizzoli.

Dickerman, Leah. 2012. "Inventing Abstraction." In *Inventing Abstraction 1910–1925: How a Radical Idea Changed Modern Art,* edited by Leah Dickerman, 12–37. New York: Museum of Modern Art.

Dickey, Tina. 1997. "Spatial Constellations: Rhythms of Nature." In *Hans Hofmann,* by Helmut Friedel and Tina Dickey, 81–91. New York: Hudson Hills Press. Originally published as *Hans Hofmann: Wunder des Rhythmus und Schönheit des Raumes.* Munich: Städtische Galerie im Lenbachhaus, 1997.

———. 2002. "A Decisive Moment: Hofmann's Mosaic Murals." In *Hans Hofmann: Revised and Expanded,* edited by James Yohe, 271–75. New York: Rizzoli.

———. 2011. *Color Creates Light: Studies with Hans Hofmann.* Victoria, BC: Trillistar Books.

———. 2012. Personal communication with the author. October 23.

Doerner, Max. 1984. *The Materials of the Artist and Their Use in Painting, with Notes on the Techniques of the Old Masters.* New York: Harcourt Brace and Co. Originally published as *Malmaterial und seine Verwendung im Bilde: Nach den Vorträgen an der Akademie der bildenden Künste in München.* Munich: Verlag für praktische Kunstwissenschaft, 1921.

Dorain, Paul B. 1958. "Electron Paramagnetic Resonance of Manganese (II) in Hexagonal Zinc Oxide and Cadmium Sulfide Single Crystals." *Physical Review* 112: 1058–60.

Eisler, Colin. 1969. "*Kunstgeschichte* American Style: A Study in Migration." In *The Intellectual Migration: Europe and America, 1930–1960,* edited by Donald Fleming and Barnard Bailyn, 544–629. Cambridge, MA: Belknap Press of Harvard University Press.

Eitner, Lorenz. 1957. "Kandinsky in Munich." *Burlington Magazine* 99 (651): 192–97, 199.

Ellsworth, Paul. 1949. "Hans Hofmann: Reply to Questionnaire and Comments on a Recent Exhibition." *Arts and Architecture* 66 (11): 22–27, 45–47.

Erlebacher, Jonah D., Eric Brown, Marion F. Mecklenburg, and Charles S. Tumosa. 1992. "The Effects of Temperature and Relative Humidity on the Mechanical Properties of Modern Painting Materials." In *Materials Issues in Art and Archaeology III: Symposium Held April 27–May 1, 1992, San Francisco, California*, edited by Pamela B. Vandiver, James R. Druzik, George Segan Wheeler, and Ian C. Freestone, 359–70. Pittsburgh: Materials Research Society.

Feinstein, Sam. 1950. *Hans Hofmann: A Film by Sam Feinstein*. Limited DVD release accompanying Patricia S. Feinstein, *Sam Feinstein*, 2008. North Truro, MA: Fields Publishing.

———. 1989. Interview by Sascha Feinstein, December. Unpublished.

———. 2008. *Portrait of Hans Hofmann as Painter, Teacher, and Friend: Reflections by Sam Feinstein*. New York: Midmarch Arts Press.

Feller, Robert L. 1964. "Critical Pigment Volume Concentration and Chalking in Paints." *Bulletin of the American Group–IIC* 5 (4): 25–26.

Feller, Robert L., and Mary Curran. 1970. "Solubility and Crosslinking Characteristics of Ethylene/Vinyl Acetate Copolymers." *Bulletin of the International Institute for Conservation–American Group* 11 (1): 42–45.

Feller, Robert L., Mary Curran, and Catherine Bailie. 1981. "Photochemical Studies of Methacrylate Coatings for the Conservation of Museum Objects." In *Photodegradation and Photostabilization of Coatings*, edited by R. H. Winslow and S. P. Pappas, 183–96. Washington, DC: American Chemical Society.

Fer, Briony. 2005. "Rothko and Repetition." In *Seeing Rothko*, edited by Glenn Phillips and Thomas Crow, 159–75. Los Angeles: Getty Research Institute.

Forma, Warren. 1963. *The Americans: Three East Coast Artists at Work*. New York: Contemporary Films.

Foster, Stephen C. 1980. *The Critics of Abstract Expressionism*. Ann Arbor, MI: UMI Research Press.

Fried, Michael. 1964. "Modern Painting and Formal Criticism." *American Scholar* 33: 642–48.

Friedel, Helmut. 1997. "To Sense the Invisible and to Be Able to Create It—That Is Art." In *Hans Hofmann*, by Helmut Friedel and Tina Dickey, 7–14. New York: Hudson Hills Press.

Gates, Glenn A., Teri Hensick, Carol Mancusi-Ungaro, Tatiana Z. Ausema, Thomas J. S. Learner, and Will Shank. 2005. "Reproducing Morris Louis Paintings to Evaluate Conservation Strategies." In *Preprints of the 14th ICOM Committee for Conservation Triennial Meeting in The Hague, 12–16 September 2005*, edited by Isabella Sourbès-Verger, 329–34. London: James and James.

Gates, Glenn A., Tatiana Z. Ausema, and Susan F. C. Lake. 2007. "What Makes the Color Field? A Technical Examination of Magna Paint." In *Modern Paints Uncovered: Proceedings from the Modern Paints Uncovered Symposium at the Tate Modern, London, May 16–19, 2006*, edited by Thomas J. S. Learner, Patricia Smithen, Jay W. Krueger, and Michael R. Schilling, 277–78. Los Angeles: Getty Conservation Institute.

GCI [Getty Conservation Institute]. 2002. "Modern Paints: A New Collaborative Research Project." *Conservation, The Getty Conservation Institute Newsletter* 17 (3): 18–20.

Geldzahler, Henry. 1972. *Hans Hofmann: The Renate Series*. New York: Metropolitan Museum of Art.

Glueck, Grace. 1977. "The 20th-Century Artists Most Admired by Other Artists." *ARTnews* 76 (9): 78–103.

Golden, Mark. 2003. Oral history interview by Joyce Hill Stoner, March 10. Foundation of the American Institute for Conservation.

Goodman, Cynthia J. 1982. "The Hans Hofmann School and Hofmann's Transmission of European Modernist Aesthetics to America." PhD diss., University of Pennsylvania.

———. 1986. *Hans Hofmann*. New York: Abbeville Press.

———. 1990. *Hans Hofmann*. New York: Whitney Museum of American Art; Munich: Prestel-Verlag.

Goodnough, Robert. 1951. "Pollock Paints a Picture." *ARTnews* 50 (3): 38–41, 60–61.

———. 1952. "Kline Paints a Picture." *ARTnews* 51 (8): 36–39, 63–64.

Greenberg, Clement. 1939. "Avant-Garde and Kitsch." *Partisan Review* 6 (5): 34–49.

———. 1945. "Art: Review of an Exhibition of Hans Hofmann and a Reconsideration of Mondrian's Theories." *Nation* 160 (16): 469.

———. 1947. "The Present Prospects of American Painting and Sculpture." *Art on the American Horizon* 93–94: 20–30.

———. 1955. "American-Type Painting." *Partisan Review* 12 (2): 179–96.

———. 1957. "New York Painting Only Yesterday." *ARTnews* 56 (4): 58–59, 84–86.

———. 1959. "Hans Hofmann: Grand Old Rebel." *ARTnews* 57 (9): 26–29, 64.

———. 1961. *Hans Hofmann*. Paris: Éditions Georges Fall.

———. 1964. *Post Painterly Abstraction: An Exhibition Organized by the Los Angeles County Museum of Art and Sponsored by the Contemporary Art Council*. Los Angeles: Los Angeles County Museum of Art.

Guston, Philip. 1990. Transcript of a 1966 public forum at Boston University with Joseph Ablow (excerpt). In *Abstract Expressionism: Creators and Critics: An Anthology*, edited by Clifford Ross, 63–75. New York: Harry N. Abrams.

Hagan, Eric, Maria Charalambides, Thomas J. S. Learner, Alison Murray, and Christina Young. 2007. "Factors Affecting the Mechanical Properties of Modern Paints." In *Modern Paints Uncovered: Proceedings from the Modern Paints Uncovered Symposium at the Tate Modern, London, May 16–19, 2006*, edited by Thomas J. S. Learner, Patricia Smithen, Jay W. Krueger, and Michael R. Schilling, 227–35. Los Angeles: Getty Conservation Institute.

Harrison, Helen A. 2007. "The Birth of Abstract Expressionism." In *Abstract Expressionism: The International Context*, edited by Joan Marter, 13–16. New Brunswick, NJ: Rutgers University Press.

———. 2014. Personal communication with the author. September 30.

Herbst, Willy, and Klaus Hunger. 2004. *Industrial Organic Pigments: Production, Properties, Applications*. Weinheim, Germany: Wiley-VCH Verlag.

Hess, Thomas B. 1953. "De Kooning Paints a Picture." *ARTnews* 52 (1): 30–33, 64–67.

———. 1965. "The Mystery of Hans Hofmann." *ARTnews* 63 (10): 39, 54–55.

———. 1970. "Editorial: Mark Rothko, 1903–1970," *ARTnews* 69 (2): 29, 66–67.

Heydenreich, Gunnar, Verena Franken, Elisabeth Jägers, Wolfgang Müller, Jenny Schultz, and Stefan Zumbühl. 2014. "Set Back the Race: Treatment Strategies for Running Oil Paint." In *Issues in Contemporary Oil Paint*, edited by Klaas Jan van den Berg, Aviva Burnstock, Matthijs de Keijzer, Jay Krueger, Thomas J. S. Learner, Alberto de Tagle, and Gunnar Heydenreich, 333–50. Amsterdam and Amersfoort: Netherlands Cultural Heritage Agency.

Hofmann, Hans. 1930a. "Art in America." *Art Digest* 4 (19): 27.

———. 1930b. Lecture at University of Minneapolis. Transcript. Hans Hofmann Archive, Bancroft Library, University of California, Berkeley. Pencil annotation: "written by H.Hm in german, translated by G. Wessels, on shipboard, on first trip to U.S."

———. 1931. "Painting and Culture." Translated by Glenn Wessels. *Fortnightly* 1 (1): 5–7.

———. 1932. "On the Aims of Art." Translated by Ernst Stolz and Glenn Wessels. *Fortnightly* 1 (13): 7–11.

———. 1932–33. "Plastic Creation." Translated by Ludwig Sander. *League* 5 (2): 11–15, 21. Reprinted in *League* 22 (3) (Winter 1950): 3–6.

———. 1938–39. "Hofmann Lecture No. VII" [typescript]. Kiesler papers, [ca. 1910]–2003. Part of the Winter 1938–39 lecture series attended by Clement Greenberg. A transcript of this lecture also appears as "Lectures on Plastic Painting, Berkely [*sic*] 1931," Archives, Special Collections, Museum of Modern Art, New York.
———. 1948. *The Search for the Real and Other Essays*, edited by Bartlett H. Hayes Jr. and Sara T. Weeks. Andover, MA: Addison Gallery of American Art.
———. 1955. "The Color Problem in Pure Painting—Its Creative Origin." In *Hans Hofmann: New Paintings, November 7–December 3, 1955*, 2–4. New York: Kootz Gallery. Reprinted frequently, including in *Arts and Architecture* 73 (2) (February 1956): 34.
———. 1959. "Statement." *It Is* 3: 10.
———. 1963a. "Hans Hofmann on Art" [transcript]. Inaugural lecture, Hopkins Center, Dartmouth College, November 17, 1962. Reproduced in *Art Journal* 22 (3) (Spring 1963): 180, 182.
———. 1963b. "The Way We See: Analysis of Eyesight in Relation to Physical Experience of Nature." *The Painter and His Problems: A Manual Dedicated to Painting*, unpublished typescript, March 21, 1963. Special Collections, Museum of Modern Art.
Hofmann, Hans, papers. [Ca. 1904]–78. Bulk 1945–65. AAA.
Hofmann, Hans, Student Dossier. 1963. Scrapbook of replies to questionnaires on Hofmann as a teacher by his students. In conjunction with the 1963–64 Museum of Modern Art exhibition *Hans Hofmann and His Students*. Special Collections, Museum of Modern Art.
Hofmann, Hans, student registries. 1918–58. Compiled by Tina Dickey for the Renate, Hans and Maria Hofmann Trust, 1991–2000. Hofmann papers, AAA.
Hughes, Gordon. 2012. "Abstraction chez Delaunay." In *Inventing Abstraction 1910–1925: How a Radical Idea Changed Modern Art*, edited by Leah Dickerman, 74–76. New York: Museum of Modern Art.
Hummelen, Ysbrand, Nathalie Menke, Daniela Petovic, Dionne Sillé, and Tatja Scholte. 1999. "Towards a Method for Artists' Interviews Related to Conservation Problems of Modern and Contemporary Art." *Preprints of the 12th ICOM Committee for Conservation Triennial Meeting in Lyon, France, 29 August–3 September 1999*, edited by Janet Bridgland, 312–17. London: James and James.
Hunter, Sam. 1956–57. "Jackson Pollock." *Bulletin of the Museum of Modern Art* 24 (2): 3–16, 18–19, 21–36.
———. 1959. *Modern American Painting and Sculpture*. New York: Dell Publishing.
———. 1964. *Hans Hofmann*. New York: Harry N. Abrams.
———. 2002. "Hans Hofmann." In *Hans Hofmann: Revised and Expanded*, edited by James Yohe, 9–39. New York: Rizzoli. A revised version of the essay that appeared in Hunter 1964.
Jaffe, Irma. 1971. "A Conversation with Hans Hofmann." *Artforum* 9 (5): 34–39.
Jewell, Edward A. 1943. "Globalism Pops into View: Puzzling Pictures in the Show by the Federation of Modern Painters and Sculptors Exemplify the Artists' Approach." *New York Times*, June 13.
Kachur, Lewis. 2007. "The View from the East: The Reception of Jackson Pollock among Japanese *Gutai* Artists." In *Abstract Expressionism: The International Context*, edited by Joan Marter, 152–62. New Brunswick, NJ: Rutgers University Press.
Kahn, Wolf. 2005. Personal communication with the author. January.
Keck, Caroline K. 1954. *How to Take Care of Your Pictures: A Primer of Practical Information*. New York: Museum of Modern Art and Brooklyn Museum.
———. 1962. *Exposition of Painting Conservation: Materials–Methods–Machines*. Brooklyn: Brooklyn Museum.
———. 1964. "History and Philosophy of Conservation." *Bulletin of the American Group, International Institute for Conservation of Historic and Artistic Works* 5 (1): 1–3.
Kiesler, Lillian. 1983. Interview by Cynthia Goodman, January 9. Kiesler papers, [ca. 1910]–2003.
Kiesler, Lillian and Frederick, papers. [Ca. 1910]–2003. Bulk 1958–2000. AAA.
Kinseher, Kathrin. 2006. "Paintings Are Made of Paint: The Exhibition of Painting Techniques in the Munich Glaspalast, 1893." In *The Object in Context, Crossing Conservation Boundaries: Contributions to the International Institute of Conservation Congress in Munich, 28 August–1 September 2006*, edited by David Saunders and Joyce Townsend, 41–48. London: IIC.
Kirby, Jo, Kate Stonor, Ashok Roy, Aviva Burnstock, Rachel Grout, and Raymond White. 2003. "Seurat's Painting Practice: Theory, Development and Technology." *National Gallery Technical Bulletin* 24: 4–37.
Klee, Paul. 1966. Transcript of a January 26, 1924, lecture. Reproduced in *Paul Klee on Modern Art*. Translated by Paul Findley. London: Faber and Faber.
Koch, Erik. 2012. Personal communication with the author. September 26.
Kootz, Samuel M. 1943. *New Frontiers in American Painting*. New York: Hastings House Publishers.
———. 1964. Oral history interview by Dorothy Seckler, April 13. AAA.
———. 1967. "The Credibility of Color. Hans Hofmann: An Area of Optimism." *Arts Magazine* 41 (4): 37–39.
Kramer, Hilton. 1966. "Symbol of Change: Hofmann, Teacher, Theorist and Artist, Codified and Passed on Modern Legacy." *New York Times*, February 18.
Krasner, Lee. 1968. Oral history interview by Dorothy Seckler, November 2. AAA.
Krauss, Rosalind E. 1994. *The Optical Unconscious*. Cambridge, MA: MIT Press.
Kremer, Georg. 2013. Personal communication with the author. December 30.
———. 2014. Personal communication with the author. January 6.
Kuh, Katherine. 1962. "Hans Hofmann." In *The Artist's Voice: Talks with Seventeen Modern Artists*, 118–29. New York: Harper and Row.
———. 2006. "Hans Hofmann in Provincetown." In *My Love Affair with Modern Art: Behind the Scenes with a Legendary Curator*, edited and completed by Avis Berman, 247–58. New York: Arcade.
Kuper, Rose, collection of Hans Hofmann miscellany. 1932–64. University of California, Berkeley, Bancroft Library.
Lake, Susan F. C. 1999. "The Relationship between Style and Technical Procedure: Willem de Kooning's Paintings of the Late 1940s and 1960s." PhD diss., University of Delaware.
Lake, Susan F. C., Suzanne Q. Lomax, and Michael R. Schilling. 1999. "A Technical Investigation of Willem de Kooning's Paintings from the 1960s and 1970s." In *Preprints of the 12th ICOM Committee for Conservation Triennial Meeting in Lyon, France, 29 August–3 September 1999*, edited by Janet Bridgland, 381–85. London: James and James.
Lake, Susan F. C., Eugena Ordonez, and Michael Schilling. 2004. "A Technical Investigation of Paints Used by Jackson Pollock in His Dripped or Poured Paintings." In *Modern Art, New Museums: Contributions to the 20th International Institute of Conservation Congress in Bilbao, 13–18 September 2004*, edited by Ashok Roy and Perry Smith, 137–41. London: IIC.
Lane, John. 1984. *Abstract Painting and Sculpture in America, 1927–1944*. New York: Harry N. Abrams.
Learner, Thomas J. S. 2000. "A Review of Synthetic Binding Media in Twentieth-Century Paints." *Conservator* 24: 96–103.
———. 2004. *Analysis of Modern Paints*. Research in Conservation Series. Los Angeles: Getty Conservation Institute.
Lippard, Lucy R., and Sam Hunter. 1967. *Ad Reinhardt: Paintings*. New York: Jewish Museum.

Lizza, Richard W. 1986. *The American Abstract Artists: Thirties' Geometric Abstraction as Precursor to Forties' Expressive Abstraction.* Ann Arbor, MI: UMI Research Press.

Lodge, Robert G. 1988. "A History of Synthetic Painting Media with Special Reference to Commercial Materials." In *American Institute for Conservation Preprints of the 16th Annual Meeting, Held in New Orleans, Louisiana, June 1–5, 1988*, 118–27. Washington, DC: AIC.

Lomax, Suzanne Q. 2005. "Phthalocyanine and Quinacridone Pigments: Their History, Properties and Use." *Reviews in Conservation* 6: 19–29.

Lomax, Suzanne Q., and Sarah L. Fisher. 1990. "An Investigation of the Removability of Naturally-Aged Synthetic Picture Varnishes." *Journal of the American Institute for Conservation* 29 (2): 181–91.

Lomax, Suzanne Q., and Thomas J. S. Learner. 2006. "A Review of the Classes, Structures, and Methods of Analysis of Synthetic Organic Pigments." *Journal of the American Institute for Conservation* 45 (2): 107–25.

Long, Rose-Carol W. 1980. *Kandinsky: The Development of an Abstract Style.* Oxford: Clarendon Press.

Maines, Christopher, Dawn Rogala, Susan F. C. Lake, and Marion F. Mecklenburg. 2011. "Deterioration in Abstract Expressionist Paintings: Analysis of Zinc Oxide Paint Layers in Works from the Collection of the Hirshhorn Museum and Sculpture Garden, Smithsonian Institution." In *Materials Issues in Art and Archaeology IX: Symposium Held November 29–December 3, 2010, Boston, Massachusetts*, edited by Pamela B. Vandiver, Chandra L. Reedy, Weidong Li, and Jose Luis Ruvalcaba Sil, 275–86. Pittsburgh: Materials Research Society.

Makela, Maria. 1996. "The Politics of Parody: Some Thoughts on the 'Modern' in Turn-of-the-Century Munich." In *Imagining Modern German Culture, 1889–1910*, edited by Françoise Forster-Hahn, 185–207. Washington, DC: National Gallery of Art.

Mancusi-Ungaro, Carol. 1990. "The Rothko Chapel: Treatment of the Black-Form Triptychs." In *Cleaning, Retouching and Coatings: Technology and Practice for Easel Paintings and Polychrome Sculpture. Preprints of the Contributions to the International Institute of Conservation Congress in Brussels, 3–7 September 1990*, edited by John S. Mills and Perry Smith, 134–37. London: IIC.

———. 2005. "Material and Method in Modern Art: A Collaborative Challenge." In *Scientific Examination of Art: Modern Techniques in Conservation and Analysis*, 152–61. Washington, DC: National Academies Press.

———. 2007. "A Sum of Corrections." In *Jasper Johns: An Allegory of Painting, 1955–1965*, edited by Jeffrey Weiss, 236–60. Washington, DC, and New Haven, CT: National Gallery of Art in association with Yale University Art Gallery.

Maor, Yonah, and Alison Murray. 2008. "Delamination of Oil Paints on Acrylic Grounds." In *Materials Issues in Art and Archaeology VIII: Symposium Held November 26–30, 2007, Boston, Massachusetts*, edited by Pamela B. Vandiver, Francesca Casadio, B. McCarthy, Robert H. Tykot, and Jose Luis Ruvalcaba Sil, 127–36. Pittsburgh: Materials Research Society.

Marontate, Janet L. A. 1996. "Synthetic Media and Modern Painting: A Case Study in the Sociology of Innovation." PhD diss., University of Montreal.

Martin, Kathleen A., Bonnie Rimer, Joseph G. Barabe, and Carol Injerd. 2013. "A New Approach to the Treatment of Fatty Acid Crystals in Oil Paintings." In *American Institute for Conservation Paintings Specialty Group, Postprints of the 38th Annual Meeting, Held in Milwaukee, Wisconsin, May 11–14, 2010*, 18–22. Washington, DC: AIC.

Matisse, Henri. 1995a. "Interview by Ernst Goldschmidt, 1911," translated by Desirée Koslin. In *Matisse on Art*, edited by Jack Flam, 60–64. Berkeley: University of California Press. Originally published as Ernst Goldschmidt, "Strejtog I Kunsten: Henri Matisse," *Politiken*, December 24, 1911.

———. 1995b. "The Role and Modalities of Colour, 1945." In *Matisse on Art*, edited by Jack Flam, 98–100. Berkeley: University of California Press. Previously published as "Rôle et modalités de la coulour." In *Les Problèmes de la peinture sous la direction de Gaston Diehl*, edited by Gaston Diehl, 237. Paris: Editions Confluences, 1945.

Mayer, Lance, and Gay Myers. 2013. *American Painters on Technique: 1860–1945.* Los Angeles: J. Paul Getty Museum.

Mayer, Ralph and Bena Frank, papers. [Ca. 1920]–64. AAA.

McClelland, Charles E. 1996. "'Young Germans, Not Young Greeks and Romans': Art, Culture, and Educational Reform in Wilhelmine Germany." In *Imagining Modern German Culture, 1889–1910*, edited by Françoise Forster-Hahn, 37–51. Washington, DC: National Gallery of Art.

McNeil, George. 1965. Oral history interview by Dorothy Seckler, June 3. AAA.

Mecklenburg, Marion, ed. 1991a. *Art in Transit: Handbook for Packing and Transporting Paintings.* Washington, DC: National Gallery of Art.

———. 1991b. *Art in Transit: Studies in the Transport of Paintings.* Washington, DC: National Gallery of Art.

Mecklenburg, Virginia M. 2008. *Modern Masters: American Abstraction at Midcentury.* London and Washington, DC: D. Giles Ltd. in association with Smithsonian American Art Museum.

Mehra, Vishwa R. 1981. "The Cold Lining of Paintings." *Conservator* 5: 12–14.

Millard, Charles. 1977. "Hans Hofmann." *Hudson Review* 30 (3): 404–8.

Miller, Bruce F. 1998. "Painting Materials Research in Munich from 1825 to 1937." In *Paint Techniques: History, Materials, and Studio Practice, Preprints from the International Institute of Conservation Congress in Dublin, 7–11 September 1998*, edited by Ashok Roy and Perry Smith, 246–48. London: IIC.

Miller, David A. 1990. "Genesis and Metamorphosis: Materials and Techniques." In *Richard Pousette-Dart*, edited by Robert Hobbs and Joanne Kuebler, 168–77. Bloomington: Indiana University Press.

MMA [Metropolitan Museum of Art]. 2012. "AXA Equitable Donates *America Today*, Thomas Hart Benton's Epic Mural Cycle Celebrating Life in 1920s America, to Metropolitan Museum." Press release, December 11.

MoMA [Museum of Modern Art]. 1965. *1963–65 Biennial Report.* New York: Museum of Modern Art.

Motherwell, Robert, Ad Reinhardt, and Bernard Karpel, eds. 1951. *Modern Artists in America.* New York: Wittenborn Schultz.

Murry, Jesse. 1981. "Hofmann's Use of Nature as Aesthetic Form." *Arts Magazine* 55 (6): 105–9.

Newman, Barnett. 1947. *The Ideographic Picture.* New York: Betty Parsons Gallery.

New York Times. 1966. "Hans Hofmann, Artist, 85, Dead. Abstract Expressionist Led a Generation of Painters through His Teachings." February 18.

O'Connor, Francis V. 1967. *Jackson Pollock.* New York: Museum of Modern Art.

O'Doherty, Brian. 1963. "Hans Hofmann: A Style of Old Age; Museum Holds Ten Year Retrospective of Veteran's Work." *New York Times*, September 15.

Osmond, Gillian. 2012. "Zinc White: A Review of Zinc Oxide Pigment Properties and Implications for Stability in Oil-Based Paintings." *Australian Institute for the Conservation of Cultural Material Bulletin* 33: 20–29.

———. 2014. "Zinc White and the Influence of Paint Composition for Stability in Oil Based Media." In *Issues in Contemporary Oil Paint*, edited by Klaas Jan van den Berg, Aviva Burnstock, Matthijs

de Keijzer, Jay Krueger, Thomas J. S. Learner, Alberto de Tagle, and Gunnar Heydenreich, 263–81. Amsterdam and Amersfoort: Netherlands Cultural Heritage Agency.
Parsons, Betty, Gallery records and personal papers. Ca. 1920–91. Bulk 1946–83. AAA.
Perls, Frank, Gallery. 1951. *The School of New York*. Beverly Hills, CA: Frank Perls Gallery.
Pomerantz, Louis. 1962. *Is Your Contemporary Painting More Temporary Than You Think? Vital Technical Information for the Present-Day Artist*. Chicago: International Book Co.
Porter, Fairfield. 1953. "Tworkov Paints a Picture." *ARTnews* 52 (3): 30–34, 72–73.
Richardson, John. 1983. "Crimes against the Cubists." *New York Review of Books* 30 (10) (June 16): 32–34. Topic continued from "Your Show of Shows," *New York Review of Books* 27 (12) (July 17, 1980): 16–24; responses printed in "Crimes against the Cubists: An Exchange," *New York Review of Books* 30 (15) (October 13, 1983): 3, and "Saving the Cubists," *New York Review of Books* 30 (17) (November 10, 1983): 60–64.
Rimer, Bonnie, Inge Fiedler, Mary A. Miller, Michael Cunningham, and Jorrit D. J. van den Berg. 1999. "Investigation of Fatty Acid Migration in Alizarin Crimson Oil Paint in Two Works by Frank Stella." In *American Institute for Conservation Paintings Specialty Group Postprints, St. Louis, Missouri, June 8–13, 1999*, 1–14. Washington, DC: AIC.
Rockwell, Tony. 1982. "Conservation Problems Present in Paintings by Hans Hofmann." Unpublished. Elise S. Haas Conservation Department, San Francisco Museum of Modern Art.
———. 2012. Personal communication with the author. November 15.
Rogala, Dawn. 2005. "Hans Hofmann from the Ground Up: Looking at the Artist's Preparatory Methods as a Window to Condition." Unpublished. Art Conservation Department, State University of New York, Buffalo.
———. 2014. "Industrial Literature as a Resource in Modern Materials Conservation: Zinc Oxide House Paint as a Case Study." In *American Institute for Conservation Paintings Specialty Group Postprints, Philadelphia, Pennsylvania, May 31–June 3, 2011*, 78–90. Washington, DC: AIC.
Rogala, Dawn, Susan F. C. Lake, Christopher Maines, and Marion F. Mecklenburg. 2010. "Condition Problems Related to Zinc Oxide Underlayers: Examination of Selected Abstract Expressionist Paintings from the Collection of the Hirshhorn Museum and Sculpture Garden, Smithsonian Institution." *Journal of the American Institute for Conservation* 49 (2): 96–113.
Rose, Barbara. 1978. "Hans Hofmann: From Expressionism to Abstraction." *Arts Magazine* 53 (3): 110–14.
———. 1980. "Jackson Pollock at Work: An Interview with Lee Krasner." *Partisan Review* 47 (1): 82–92.
Rosenberg, Harold. 1952. "The American Action Painters." *ARTnews* 51 (8): 22–23, 48–50.
———. 1957. "Hans Hofmann: Nature into Action." *ARTnews* 56 (3): 34–36, 55–56.
———. 1963a. "The New as Value." *New Yorker* 39 (29): 136–46.
———. 1963b. "Hans Hofmann and the Stability of the New." *New Yorker* 39 (37): 100, 103–5, 108–10.
———. 1966. "Hans Hofmann, 1880–1966." Eulogy delivered at Hofmann's funeral, February 20. *ARTnews* 65 (2): 21.
———. 1969. "Art and Words." *New Yorker*, March 29.
———. 1972. Oral history interview by Paul Cummings, April 7. AAA.
———. 1975. "Criticism and Its Premises" [transcript]. Seminar in Art Education for Research and Curriculum Development, Pennsylvania State University, 1966. Printed in *Art on the Edge: Creators and Situations*, 135–52. New York: Macmillan.
———. 1982. *The Anxious Object: Art Today and Its Audience*. Chicago: University of Chicago Press. Originally published by Horizon Press, 1964.
Rosenthal, Lou. 2005. Personal communication with the author.
Rothko, Mark. 1951. "Statement." *Interiors* 110 (10) (May): 104. Excerpt of Rothko's statement at the March 19, 1951, Museum of Modern Art symposium "How to Combine Architecture, Painting, and Sculpture."
Rowell, Margit. 1980. *Ad Reinhardt and Color*. New York: Solomon R. Guggenheim Museum.
Sander, Ludwig. 1969. Oral history interview by Paul Cummings, February 4–12. AAA.
Sandler, Irving. 1963. "In the Art Galleries." *New York Post Magazine*, Sept. 15.
———. 1970. *The Triumph of American Painting: A History of Abstract Expressionism*. New York: Praeger.
———. 1973. "Hans Hofmann: The Pedagogical Master." *Art in America* 61 (3): 48–57.
———. 1978. *The New York School: The Painters and Sculptors of the Fifties*. New York: Harper and Row.
———. 2009. *Abstract Expressionism and the American Experience: A Reevaluation*. New York: Hudson Hills Press.
———. 2012. Personal communication with the author. August 27.
Schmutz, F. C. 1935. "Primers for Exterior House Paints." *Official Digest—Federation of Paint and Varnish Production Clubs* 141: 355–57.
Seckler, Dorothy G. 1953. "Stuart Davis Paints a Picture." *ARTnews* 52 (4): 30–33, 73–74.
———. 1977. *Provincetown Painters: 1890s–1970s*. Syracuse, NY: Everson Museum of Art.
Seitz, William C. 1955. "Abstract-Expressionist Painting in America: An Interpretation Based on the Work and Thought of Six Key Figures." PhD diss., Princeton University. Published as *Abstract Expressionist Painting in America*. Cambridge, MA: Harvard University Press, 1983.
———. 1963. *Hans Hofmann: With Selected Writings by the Artist*. New York: Museum of Modern Art.
Selz, Peter. 1993. "Hans Hofmann: Selections from the Artist's Gift to the University." *American Art Review* 5 (2): 126–31.
———. 1997. *German Expressionist Painting*. Berkeley: University of California Press. Originally published in 1957. Book based on his 1954 University of Chicago dissertation on the same topic.
———. 2002. "Forming with Color." In *Hofmann: Evolution/Revolution*, 5–6. San Francisco: Hackett-Freedman Gallery.
SFMOMA [San Francisco Museum of Modern Art] records. 1966–present. Unpublished records related to the conservation of Hofmann works in the collection of BAM/PFA; includes information related to the conservation of works prior to museum acquisition. Records of the Elise S. Haas Conservation Department, SFMOMA.
Shank, William J. 1991. "An Overview of the Structure and Condition of the Hans Hofmann Paintings, University Art Museum, Berkeley, California." Elise S. Haas Conservation Department, San Francisco Museum of Modern Art.
Shiff, Richard. 1987. "Performing an Appearance: On the Surface of Abstract Expressionism." In *Abstract Expressionism: The Critical Developments*, 94–123.
———. 2011. *Between Sense and de Kooning*. London: Reaktion Books.
Simon, Sidney. 1967. "Concerning the Beginnings of The New York School: 1939–1943 (Interviews with Busa, Matta and Motherwell, Conducted by Sidney Simon)." *Art International* 11 (6): 17–23.
Singer, Sandra L. 2003. *Adventures Abroad: North American Women at German-Speaking Universities, 1868–1915*. Westport, CT, and London: Praeger.
Spoerri, Max. 2005. Personal communication with the author. January.
———. 2011. Personal communication with the author. October 19.

Standeven, Harriet A. L. 2003. "The Historical and Technical Development of Gloss House Paints, with Reference to Their Use by Twentieth-Century Artists." PhD diss., Royal College of Art.

———. 2007. "'Cover the Earth': A History of the Manufacture of Household Gloss Paints in Britain and the United States from the 1920s to the 1950s." In *Modern Paints Uncovered: Proceedings from the Modern Paints Uncovered Symposium at the Tate Modern, London, May 16–19, 2006,* edited by Thomas J. S. Learner, Patricia Smithen, Jay W. Krueger, and Michael R. Schilling, 75–82. Los Angeles: Getty Conservation Institute.

Stedelijk Museum Amsterdam. 2002. *Barnett Newman: Cathedra.* Amsterdam: Stedelijk Museum Amsterdam.

Stella, Frank. 1999. "The Artist of the Century." *American Heritage* 50 (7): 14–17.

Still, Clyfford. 1935. "Cézanne: A Study in Evaluation." MFA thesis, State College of Washington.

Stoner, Joyce H. 2004. "America's Colormen: Bocour, Levison, Gamblin, and Golden." In *Modern Art, New Museums: Contributions to the 20th International Institute of Conservation Congress in Bilbao, 13–18 September 2004*, edited by Ashok Roy and Perry Smith, 189–92. London: IIC.

Stout, George L. 1948. *The Care of Pictures.* New York: Columbia University Press.

Stringari, Carol, Ellen Pratt, and Christopher McGlinchey. 2004. "Reversal Versus Retirement: Study and Treatment of *Black Painting,* 1960–66 by Ad Reinhardt." In *Modern Art, New Museums: Contributions to the 20th International Institute of Conservation Congress in Bilbao, 13–18 September 2004*, edited by Ashok Roy and Perry Smith, 165–69. London: IIC.

Tallent, Carolyn. 1987. "Notes from One Hour Interview with Erle Loran," June. Unpublished.

———. 1988. "Investigation of the Painting Materials and Techniques of Hans Hofmann—Preliminary Report." Unpublished fellowship report, August 1. Intermuseum Conservation Association. Report files include a handwritten chart compiling data.

———. 2005. Personal communication with the author. January 23.

Time [magazine]. 1956. "The Age of Experiment." *Time* 67 (7): 62–67.

Toch, Maximilian. 1925. *The Chemistry and Technology of Paints.* New York: D. Van Nostrand.

Van Bommel, Maarten, Muriel Geldof, and Ella Hendricks. 2005. "An Investigation of Organic Red Pigments Used in Paintings by Vincent van Gogh (November 1885 to February 1888)." *ArtMatters: Netherlands Technical Studies in Art* 3: 111–37.

Varnedoe, Kirk. 1998. "Comet: Jackson Pollock's Life and Work." In *Jackson Pollock: New Approaches*, edited by Kirk Varnedoe and Pepe Karmel. New York: Museum of Modern Art.

Villers, Caroline, ed. 1974. *Conference on Comparative Lining Techniques.* Greenwich, UK: National Maritime Museum.

Watherston, Margaret M. 1974. "Report on the Conservation of Recent Paintings: The Cleaning of Color Field Paintings." In *The Great Decade of American Abstraction: Modernist Art, 1960 to 1970*, edited by E. A. Carmean Jr., 119–29. Houston: Museum of Fine Arts. Reprint of paper presented at the annual meeting of the International Institute for Conservation–American Group in Oberlin, Ohio, on June 5, 1971, and again at the Fifth International Institute for Conservation Congress, held in Lisbon, October 9–14, 1972.

Weldon, Dwight G. 2001. *Failure Analysis of Paints and Coatings.* New York: John Wiley and Sons.

Werthan, S. 1947. "Post-war Exterior House Paints." *Paint, Oil and Chemical Review* 110: 37–39.

Wessels, Glenn. 1967. Oral history interview by Suzanne B. Riess. University of California, Berkeley, Regional Oral History Office.

Wight, Frederick S. 1957. *A Retrospective Exhibition of Hans Hofmann.* Berkeley and Los Angeles: University of California Press.

Wijnberg, Louise, Elisabeth Bracht, Klaas Jan van den Berg, and Matthijs de Keijzer. 2011. "A Study of the Grounds Used by Three Post-War American Artists (1954–1975): Barnett Newman, Ellsworth Kelly and Brice Marden." CD-ROM. In *Preprints of the 16th ICOM Committee for Conservation Triennial Meeting in Lisbon, Portugal, 19–23 September 2011*, edited by Janet Bridgland, 10 pages. London: James and James.

Wimsat, Justine S. 1982. "Wax Emulsion, Tempera, or Oil? Arthur Dove's Materials, Techniques, and Surface Effects." In *AIC Preprints, Papers Presented at the Tenth Annual Meeting of the American Institute for Conservation of Historic and Artistic Works, Milwaukee, Wisconsin, May 26–30, 1982*, 183–88. Washington, DC: AIC.

Wolanin, Barbara. 2003. "Prelude to Provincetown: Hofmann, Matter, Carles, and the Legacy of Their 1934 Gloucester Summer." In *From Hawthorne to Hofmann: Provincetown Vignettes, 1899–1945*, edited by Richard J. Boyle, 40–57. New York: Hollis Taggart Galleries.

Wolf, Ben. 1945. "The Digest Interviews Hans Hofmann." *Art Digest* 19: 52.

Yohe, James. 2013. Personal communication with the author. December 4.

Young, Christina, Rebecca Gregg, Roger Hibberd, James Walker, and Thomas J. S. Learner. 2004. "The Physical Properties of Modern Commercially Available Primings and Their Interaction with Subsequent Paint Layers." In *Modern Art, New Museums: Contributions to the 20th International Institute of Conservation Congress in Bilbao, 13–18 September 2004*, edited by Ashok Roy and Perry Smith, 244. London: IIC.

Illustration Credits

All Hans Hofmann works are reproduced with permission of the Renate, Hans & Maria Hofmann Trust and the University of California, Berkeley Art Museum and Pacific Film Archive / Artists Rights Society (ARS), New York.

Frontispiece, figs. 3.16–3.17: © 2016 Estate of Rudy Burckhardt / Artists Rights Society (ARS), New York / Courtesy Ameringer McEnery Yohe

P. x: Andreas Feininger / The LIFE Picture Collection / Getty Images

Fig. 1.1: Städtische Galerie im Lenbachhaus und Kunstbau, Munich

Figs. 1.2–1.5, 1.9, 2.5, 2.7, 3.1–3.2: Courtesy Ameringer McEnery Yohe

Fig. 1.6: Solomon R. Guggenheim Museum, New York

Fig. 1.7: Archives of American Art, Smithsonian Institution

Fig. 1.8: Courtesy of The Bancroft Library, University of California, Berkeley, BANC MSS 80/27 c box 1 folder 1

Figs. 2.1, 3.4, 3.7–3.11, 3.14, 4.7, 6.3, 6.8–10: University of California, Berkeley Art Museum and Pacific Film Archive / Photo: Benjamin Blackwell

Figs. 2.2, 2.4, 3.5, 6.5: Digital Image © The Museum of Modern Art / Licensed by SCALA / Art Resource, NY

Fig. 2.3: © 2016 Museum of Fine Arts, Boston

Figs. 2.6, 3.3: Arnold Newman / Getty Images

Fig. 2.8: © A. E. Artworks, LLC

Figs. 3.6, 5.3, 5.8, 6.13–6.14: Memorial Art Gallery of the University of Rochester

Figs. 3.12–3.13: Albright-Knox Art Gallery / Art Resource, NY

Fig. 3.15: Fred W. McDarrah / Getty Images

Figs. 4.2–4.6, 4.8–4.13, 5.2, 5.5–5.6, 5.9, 6.2, 6.4, 6.6–6.7, 6.11–6.12: Dawn Rogala

Fig. 6.15: Estate of Herbert Matter / Special Collections and University Archives Department, Stanford University Libraries

Index

Acknowledgments

I present this book as a thank-you to the many people and institutions that have supported my work and my journey, and as a testament to their generosity and guidance. I was introduced to the work of Hans Hofmann during Bill Agee's research methods seminar years ago at Hunter College of the City University of New York, and I have been studying Hofmann ever since. I soon found my way to the Renate, Hans and Maria Hofmann Trust and to a research position on the Hofmann catalogue raisonné project. The study of Hofmann's materials that I began as a conservation student at Buffalo State College of the State University of New York led to postgraduate research on the Abstract Expressionist collections of the Smithsonian Institution, and to my doctoral research undertaken at the University of Delaware. *Hans Hofmann: The Artist's Materials* is based on my dissertation. "Hofmann offers a lesson in patience," said Clement Greenberg. "That lesson, from him and from others, I shall never finish learning" (Greenberg 1961, 58). The exclusion of any names from the list below is an unintentional oversight on my part and in no way lessens any individual's role in this project or my gratitude for his or her invaluable assistance.

My thanks go to my dissertation committee: Joyce Hill Stoner (chair), Murray Johnston, Christopher Maines, Jill Sterrett, and Roberta Tarbell. I owe an equal debt to the current and former trustees, staff, and colleagues of the Hofmann Trust for their unwavering support—especially Patricia Gallagher, John Powers, and Robert Warshaw; collections manager Stacey Gershon; and Ameringer McEnery Yohe gallery director James Yohe.

My research would not have been possible without the encouragement and collaboration of staff at numerous art and research institutions including registrar Laura Fleischmann, Albright-Knox Art Gallery; my colleagues at the Museum Conservation Institute, Smithsonian Institution, especially director Robert Koestler, deputy director Paula DePriest, and research scientist emeritus Marion Mecklenburg; assistant curator Jessica Marten and former chief curator Marjorie Searl, Memorial Art Gallery, University of Rochester; conservation staff at the Museum of Modern Art, especially chief conservator James Coddington and paintings conservators Michael Duffy and Anny Aviram; conservation division staff at the National Gallery of Art, Washington, DC, especially conservation scientist Chris Maines and head of painting conservation Jay Krueger; collections care staff at the San Francisco Museum of Modern Art, especially director of collections and conservation Jill Sterrett and conservators Paula De Cristofaro and Alina Remba; and curatorial and collections staff at the University of California, Berkeley Art Museum and Pacific Film Archive, especially chief curator Lucinda Barnes and director of registration Lisa Calden. Others who played key roles include Stuart Croll, Mark Golden, James Hamm, Georg Kremer, Susan Lake, and Debbie Hess Norris.

My sincere thanks go to those who shared their time, memories, and research, including Tina Dickey, Pat Stark and Sascha Feinstein, Wolf Kahn, Erik Koch, Robert Lodge, Tony Rockwell, Lou Rosenthal, Irving Sandler, Richard Cándida-Smith, Max Spoerri, and Carolyn Tallent.

My research was made possible by generous financial support, including a Paul Coremans Endowment Art Conservation Research Fellowship and UD College of Arts and Sciences Dean's Doctoral Scholarship, a Daughters of the American Revolution Centennial Scholarship, and a publication grant from the Society for the Preservation of American Modernists. A fortuitously timed Save America's Treasures grant to the University of California, Berkeley Art Museum and Pacific Film Archive brought the bulk of its Hofmann paintings out of long-term storage for conservation assessment and thereby provided a brief window of unprecedented access to this collection.

I would also like to thank the peer reviewers, whose comments and suggestions helped streamline my text and mix of topics. And lastly, this book would not exist without the support and hard work of many people at the J. Paul Getty Trust. At the GCI, my sincere thanks go to head of science Tom Learner, for his belief in my work, and to publications manager Cynthia Godlewski, for supporting the publication of my research as part of the GCI's Artist's Materials series. At Getty Publications, I extend my gratitude to editor Tevvy Ball, for continually raising my spirits and productivity, and to the rest of the team, especially production coordinator Suzanne Watson, designer Jeffrey Cohen, and image rights coordinators Pam Moffat and Pauline Lopez. Their vision and talent made possible this contribution to the technical study of modern art.

Dawn V. Rogala

About the Author

DAWN V. ROGALA graduated from the MA/CAS program in art conservation at Buffalo State College of the State University of New York and received her PhD in preservation studies from the University of Delaware. She has held positions or trained at museums and research facilities in the United States and Europe, including the Isabella Stewart Gardner Museum, the Smithsonian Institution, Yale University, the Wadsworth Atheneum, and the Stichting Restauratie Atelier Limburg. Rogala has authored and coauthored papers on materials behavior, paint analysis, and research methodology, and she has spent the past fifteen years studying and treating the works of Hans Hofmann. Her conservation of modernist paintings includes the treatment of works by such Hofmann contemporaries as Franz Kline, Robert Motherwell, and Jackson Pollock, as well as mixed-media paintings by Suzanne Duchamp-Crotti and her brother Jacques Villon in the collections of the Société Anonyme; and works by CoBrA cofounder Karel Appel, El Paso group cofounder Antonio Saura, and Washington Color School artists Leon Berkowitz, Gene Davis, and Kenneth Noland. Rogala is currently a paintings conservator at the Museum Conservation Institute of the Smithsonian Institution, where she pursues the conservation and technical study of paintings and paint materials, with a focus on modern and contemporary artworks and artists' materials.